AF651232

Environmental Biotechnology

Environmental Biotechnology

Hemant Rawat

Oxford Book Company
Jaipur ▪ New Delhi

Published By

Oxford Book Company

Regd. Off. "Bony Residency", Gate No. 2
Opp. Tilak Public School, Vishweswariya Nagar,
Gopalpura Road, Jaipur - 302018, India
Ph: 0141 - 2761280, Telefax: 2761381
email - oxfordbookjpr@gmail.com
website - oxfordbookcompany.com

Delhi Office: 102, 1st Floor, Satyam House,
4327/3, Ansari Road, Darya Ganj,
New Delhi - 110002, India
Ph: 011-45652440

Distributed By

KEYA INTERNATIONAL PRESS LLP
E-mail: keyainternationalpressllp@gmail.com

Published in India, 2025

© All Rights Reserved

ISBN : 9789355248930

Price: ₹ 2195.00

All Rights Reserved. No part of this publication may be reproduced, stored in a retrieval system, or transmitted, in any form or by any means, electronic, mechanical, photocopying, recording, scanning or otherwise, without the prior written permission of the copyright owner.

Typeset by Shivangi Computers, Jaipur

Printed and bound in India by Thomson Press Ltd.,
New Delhi

Preface

The complexity of the ethical dilemmas posed by biotechnology is such that we are still trying to identify what questions should be asked. The book recognizes that the first step in making the types of ethical decisions which are increasingly demanded of us is searching for the right questions. The chapters vary greatly in both length and orientation.

The advent of new technologies has resulted in extensive exploitation of the environment causing serious environmental problems. It has become very important to adopt clean technologies for pollution prevention as well as modern technologies for remediation and management of environment.

Biotechnology has emerged as a powerful and valuable tool for employing clean technologies and for sustainable environmental management. This will make it useable by a wide range of groups — from students in the classroom, to professionals such as scientists, medical doctors, business people, lawyers, and ethicists, as well as environmental and other socially active groups.

What becomes obvious from the chapters here is that the issues arising from biotechnology are dramatic and will impact all of us and thus are in need of wide-ranging public debate. It is hoped that this book will contribute to that debate.

Hemant Rawat

Contents

	Preface	*vii*
1.	Introduction	1
2.	Biodiversity Value and Loss	40
3.	Biotechnology and Communication	84
4.	Environmental Applications of Biotechnology	118
5.	The Environmental Treaty	157
6.	Biotechnology with Green Credentials	197
7.	Biotchnology vs Ethics and Politics	248
	Index	297

1

Introduction

NEXT GENERATION TECHNOLOGIES

Traditionally, technology has addressed environmental problems by improving "end-of-pipe" controls, that is, by providing better means to control the release of hazardous substances or to clean up subsequent releases. More recently, technology has been used to enhance pollution prevention, by limiting the use or creation of environmentally hazardous materials. NGETs contribute to this effort by focusing on the redesign, at a molecular level, of manufacturing processes and products. Because such redesign provides other benefits to firms, including increases in production efficiency and reductions in the requirements for raw materials, NGETs may in some cases offer the prospect of environmental protection at no additional manufacturing cost

The class of NGETs has significant overlap with the emerging scientific discipline of green chemistry. Green chemistry focuses on the design of chemical processes and products that are environmentally benign. It is guided by a set of 12 principles. The term *green chemistry* is being used ever more broadly—describing a programme within the U.S. Environmental Protection Agency (EPA), designating programmes or centers in a number of universities, and providing the name of the Green Chemistry Institute, a partnership of a network of scientists and institutions with the American Chemical Society.

Any technology resulting from green chemistry research most likely qualifies as an NGET, and because most environmental hazards stem from chemicals and their effects on society, green chemistry may provide a significant fraction of all NGETs. These technologies are the focus of our review. However, we do not exclude a few technologies dealing with physical processes—such as joining, finishing, or magnetic cooling— that stem from other areas of scientific research. Thus this report focuses on technologies from green chemistry but describes its subject with the more general term: "Next Generation Environmental Technologies."

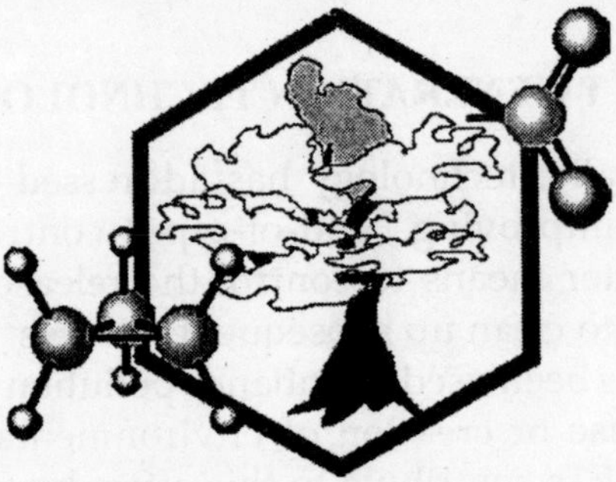

Fig. Green Chemistry

Green chemistry is aimed to help practitioners involved in the invention, design, and application of chemical products and processes to reduce or eliminate the use and generation of hazardous substances. The principles promote environmental chemistry at all levels: research, reduction to practice, education, national and international policy, and public perception. They help chemists and chemical engineers design more environmentally benign products and processes through the selection of feedstocks, reagents, alternative synthetic transformations, solvents, reaction conditions, and end products, as well as the design of safer chemicals.

For instance, when considering what feedstocks to use in generating a particular compound, the green chemist will explore renewable feedstocks whenever practical. There is no guarantee that such renewable feedstocks are possible for a given reaction nor, if they are, that they will provide a net environmental benefit. Nonetheless, the principles of green chemistry provide a set of design criteria and goals that can help improve the environmental performance of new products and processes.

As they begin to apply the principles of green chemistry, chemists are discovering many pathways toward environmentally greener designs. For instance, synthetic catalysts and biocatalysts such as enzymes offer a variety of alternative synthetic pathways that are consistent with the principles of green chemistry.

These catalysts also enable the use of alternative reagents and feedstocks, previously impractical with conventional chemistries. The two crosscutting enabling technologies hold much promise in fulfilling the expectations for green chemistry.In surveying the field, it is useful to employ a taxonomy of methods that develop NGETs. To that end, we use the seven areas of green chemistry, a taxonomy that has been laid out to help describe green chemistry research:

1. Use of alternative feedstocks that are both renewable and less toxic to human health and to the environment.
2. Use of innocuous reagents that are inherently less hazardous and are catalytic.
3. Employment of natural processes— biosynthesis, biocatalysis, and biotech-based chemical transformations for both efficiency and selectivity.
4. Use of alternative solvents that reduce potential harm to the environment and serve as alternatives to currently used volatile organic solvents, chlorinated solvents, and other hazardous chemicals.
5. Safer chemical design— with principles of toxicology to minimize intrinsic hazards while maintaining needed functionality.
6. Development of alternative reaction conditions that increase selectivity and enable easier separations.
7. Minimization of energy consumption.

All the case studies considered in this report employ one or more of these methods.

ASSESSING THE PROMISE OF NGETS

• NGET proponents foresee a promising future based on an increasing awareness of green chemistry and growth in commercial applications over the next two decades. The authors of the chemical industry's *Vision 2020* are typical in their optimism:

> While the challenges of sustainability are significant, there are also major opportunities. The chemical industry now has the opportunity to accelerate its development of advanced manufacturing technologies and new chemistry and related technologies that use materials and energy more efficiently. U.S. companies also have an opportunity to build on their current dominance in the relatively new field of environmental technology. Environmental technologies make sustainable development possible by reducing risk, improving process efficiency, and creating products and processes that are environmentally beneficial or benign.
>
> To support this vision, the chemical industry has established goals of reducing feedstock losses to waste and by-products by 90 percent, energy intensity by 30 percent, and emissions and effluents by 30 percent, all by the year 2020.

This report provides a first step toward assessing the potential of NGETs to help meet such goals identified by the chemical and many other industries. We examined 25 existing NGETs in this process. The technologies considered span the range of development from early research to full deployment in profitable businesses. Those that have been commercialized are offering significant benefits. Others that are still in various developmental stages offer similar promise.

In addition to supporting more public and private sector research, policymakers will be considering appropriate mechanisms to express society's demand for the benefits that NGETs can offer. Even when NGETs can provide such benefits at little or no cost, they must compete inside firms with other investment opportunities. Furthermore, the diversity of potential NGET applications offers both a benefit and a challenge. Policymakers are unlikely to know beforehand the full range of future benefits that are possible. Thus for NGETs to play their full role in a new approach to environmental protection, policymakers will need to determine society's demand for such benefits, the best ways to achieve them, and the rationale for actions to be taken, communicating these to industry and society at large.

Twelve Principles of Green Chemistry

1. It is better to prevent waste than to treat or clean up waste after it is formed.
2. Synthetic methods should be designed to maximize the incorporation of all materials used in the process into the final product.
3. Wherever practicable, synthetic methodologies should be designed to use and generate substances that possess little or no toxicity to human health and the environment.
4. Chemical products should be designed to preserve efficiency of function while reducing toxicity.
5. The use of auxiliary substances (e.g., solvents, separation agents) should be made unnecessary wherever possible and innocuous when used.
6. Energy requirements should be recognized for their environmental and economic impacts and should be minimized. Synthetic methods should be conducted at ambient temperature and pressure.
7. A raw material of feedstock should be renewable rather depleting wherever technically and economically practicable.
8. Unnecessary derivitization (blocking group, protection/ deprotection, temporary modification of physical/ chemical processes) should be avoided whenever possible.

9. Catalytic reagents (as selective as possible) are superior to stoichiometric reagents.
10. Chemical products should be designed so that at the end of their function they do not persist in the environment and break down into innocuous degradation products.
11. Analytical methodologies need to be further developed to allow for real-time, in-process monitoring and control prior to the formation of hazardous substances.
12. Substances and the form of a substance used in a chemical process should be chosen so as to minimize the potential for chemical accidents, including releases, explosions, and fires.

Gaining the Benefits from NGETs

The case studies demonstrate that NGETs offer potentially significant benefits over a wide variety of applications, recognizing the uncertainty in any forecast of the ultimate range and scope for any specific technology. This report describes current benefits and then the possible longer-term benefits that depend on further technology advances. We now look at the mechanisms by which the benefits are delivered.

Mechanisms

A review of our case studies and the literature on environmental technologies suggests that the invention, design, and deployment of NGETs offer four types of benefits—environmental, economic, security, and health and safety. These benefits are delivered by three mechanisms: cost reduction, risk reduction, and broad technology transfer or spillover effects.

Cost Reduction

NGETs can reduce the costs of providing the various benefits. The classic model derived from welfare economics provides a way of thinking about how NGETs might enable a new approach to environmental protection. In this model, one can estimate the costs associated with emitting some pollutants. One also knows the potential costs of reducing these emissions. With this information, policymakers can choose the level of environmental regulation so that the marginal cost of reducing emissions is just balanced by the marginal benefit of further reductions.

At this optimal level of reductions, society as a whole has gained a benefit equal to the reduced level of environmental impact less the cost of achieving the necessary reductions. In this model, technological innovation can play a role by lowering the costs of reducing emissions. Policymakers are then able to choose a higher level of regulation, with an overall increase in the net societal gain.

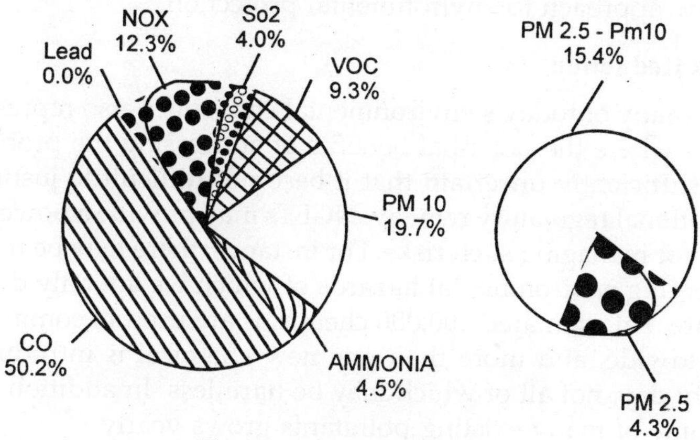

Fig. Emission of pollutant

In this model, such research can play its most important role when the current costs of reducing some pollutant are sufficiently high that the optimal level of regulation leaves the pollutant untouched. An opportunity is created for research to make significant cost reductions compared to the alternatives. In cases where costs are already low compared with benefits, innovation may provide little additional gain. Only in cases where innovation makes dramatic improvements in relative costs can it fundamentally change the regulatory calculus. Many of today's environmental challenges represent cases where the current costs of reductions are high compared with the perceived benefits.

One example might be the sequestration of greenhouse gases from electric power generation facilities, where current costs may be high because of the limited or untested technological options available for reductions. NGETs may significantly change the situation by providing technological options that dramatically reduce the costs of reductions relative to the derived social benefits. In such cases the government can set new regulatory standards and achieve previously unavailable social benefits. NGETs may also significantly change the situation if they can lower costs of reductions sufficiently so that firms will make these reductions without the need for traditional regulatory stimulus. This latter case represents the situation where NGETs can indeed provide a new approach to environmental protection.

Risk Reduction

Many of today's environmental challenges also represent cases where the potential benefits of addressing the problem are sufficiently uncertain that it becomes difficult to justify a traditional regulatory remedy. NGETs may provide a powerful tool for managing such risks. For instance, there may be many emerging environmental hazards of which we are only dimly aware. An estimated 100,000 chemicals are used in commerce worldwide, and more than one new chemical is introduced every day, not all of which may be harmless. In addition, the volume of many existing pollutants grows yearly.

The "precautionary principle" represents one approach to addressing such uncertain environmental risks. This principle, embodied in many of the environmental policies of the European Union, suggests that human activities with a potential for serious environmental harm should be limited even in the absence of full scientific certainty. For instance, the principle suggests that some new chemicals ought not be used until proved completely safe.

Industry and some policymakers in the United States generally oppose this concept because it can increase the risks to economic growth in exchange for an uncertain benefit. Such growth can create additional wealth that can help pay for enhanced environmental protection, and is a valuable end in itself. To preserve economic growth, many experts argue that hazardous substances should not be regulated until the benefits of doing so clearly outweigh the costs. However, some environmental dangers may not become certain until damages are already significant and emissions control and cleanup have become expensive.

NGETs may help serve as a bridge between these two positions by making a different wager. NGETs can reduce environmental risk by eliminating broad classes of potentially dangerous substances from products and processes. The principles of green chemistry guide firms in designing new products and processes in such a way that their impact on the environment is reduced. While reductions of many new effluents will have little ultimate consequence, NGETs may unknowingly eliminate some critical environmental problems before we ever learn that such problems exist.

Technology Transfer

Finally, research on NGETs can have broad spillover effects that facilitate the development of subsequent NGETs in the same firm, or across an industry. It is widely understood that society's scientific and technological advances provide a stock of knowledge that can be repeatedly drawn on by those seeking new innovations. Early NGETs may have had a relatively sparse base on which to draw. As the body of research and practical experience grows, it will become easier and less expensive for firms to make progress in their particular applications.

In addition, innovation in new technology areas can be subject to increasing returns. Scientists and engineers can draw more heavily on the principles and lessons of many NGETs and green chemistries as they become more familiar and proven. Constraints may ease as NGETs become more prevalent. Early research efforts can help create the knowledge and resources to make future projects easier and more likely to succeed. Spillover effects also occur when one firm installs technology eliminating a particular pollutant. Competitors may then be pressured through a variety of forces to at least reduce, if not eliminate, that pollutant.

The importance of spillover effects for the development of NGETs is increased by the wide variety of fields from which these technologies derive. Disciplines other than those shown can also play a role. These illustrative connections were generated by the authors, observing patterns of citations in the chemical sciences literature and individual histories of the developments in our case studies. This diversity underlies the great promise of NGETs, but it also stresses the difficulty of bringing together all the potentially relevant information while these technologies and their use remain relatively new and unexplored.

Near-Term Benefits of NGETs

Proponents for NGETs and associated green chemistry research suggest that this new approach to environmental protection will have major impacts in the decades ahead. We use our case studies to see to what extent this may be true. In particular, we examine in the case studies how NGETs, to date, have helped reduce the use of hazardous substances, the benefits NGETs provide, why NGETs are adopted, and the barriers to their greater adoption.

A key challenge in such an assessment is that NGETs represent new technology. The federal R&D support for green chemistry programmes is only about a decade old, relatively young to have had significant societal impact. Accordingly, two-thirds of our case studies represent technologies not yet embodied in sustainable businesses. Despite their infancy, our case studies demonstrate that NGETs have already provided important societal benefits.

Our case studies demonstrate that NGETs can provide benefits across the full spectrum considered in this study—environmental, security, occupational safety and health benefits, and economic. The vast majority of our case studies primarily offer environmental or economic benefits. This distribution owes in part to our selection, which focused on NGETs that offer environmental benefits at low or no cost.

However, this distribution of cases also reflects the significant interest over the past decade in developing new technologies to enhance environmental protection. The next decade may also see increased technological activity directed toward national security.

Environmental Benefits

Environmental benefits from NGETs arise from a variety of techniques that either reduce or eliminate the use of hazardous materials, or replace environmentally sensitive materials with lessdamaging chemicals. NGETs offer improved environmental quality in a range of areas, including reduced resource depletion, ecosystem damage, ozone depletion, and levels of toxic pollutants, as well as greater material efficiency, waste minimization, and climate improvements.

As the use of different chemicals has expanded to include many thousands of specialty chemicals across the economy, so too have the means of tracking these chemicals. However, this is done with some difficulty. The Emergency Planning and Community Right-to-Know Act (EPCRA) was enacted in 1986 as a result of pressure from public and environmental organizations' demands to know what chemicals were being released into communities without their knowledge.

EPCRA Section 313 requires EPA and the states to collect annually information on a list of chemicals that industrial facilities may be releasing into the environment. These data, known as the Toxics Release Inventory (TRI), are collected on only a small portion (about 650 chemicals) of the total number of chemicals used in society. However, the Inventory may be used as a barometer for the fate of some of the more commonly used chemicals and a gauge for understanding the use of some of the more dangerous and highly pervasive materials.

The number of chemicals for which data are compiled each year has increased from 343 chemicals in 1994 to more than 650 chemicals in 30 categories today. The top six two-digit Standard Industrial Classification (SIC) code industries comprise more than 92 percent of the total releases.

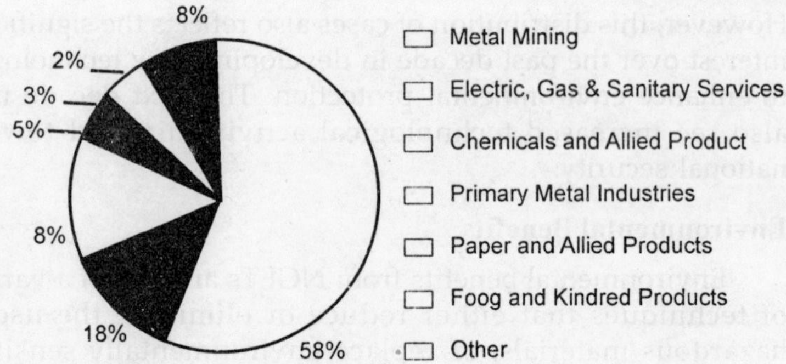

Fig. Breakdown of Total TRI Chemical Releases Over All Industries.

The mining industry (SIC codes 10 and 12) releases the bulk (56 percent) of the TRI chemicals, nearly 4 billion pounds of the total 7.1 billion pounds, and almost all (99 percent) is released on site. Only a couple of case studies focused on the mining industry, and none of them directly addressed the largest problem there, which is the removal of underground material left on the surface.

Paper and allied products releases 18 percent, and primary metals 5 percent. The chemicals and allied products industry (SIC code 28) releases approximately 600 million pounds "Toxics Release Inventory (TRI) Programme." 22 of chemicals (8 percent) into surface water, over land, underground, and into the air. These compounds are widely dispersed and the top ten chemicals account for only 64 percent of total releases. The bulk of the case studies mentioned in the report address chemicals used in this industry sector.

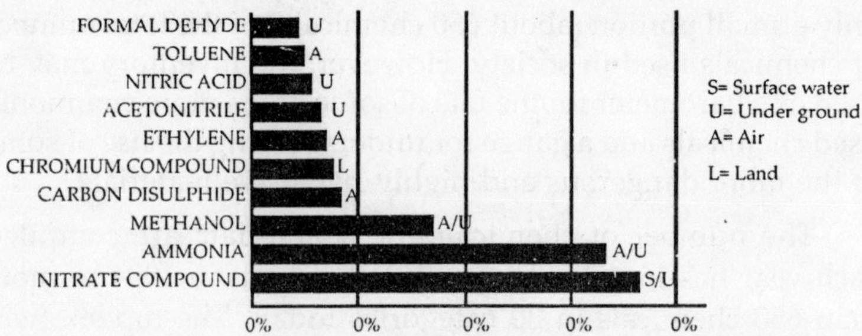

Fig. TRI Release Breakdown of the Chemical and Allied Products Industry (SIC code 28) (per cent of emissions)

The top ten chemicals released into the air across all industries are largely comprised of solvents and cover 77 percent of the total released into the air. The top ten chemicals released from all industries excluding mining. These comprise nearly 3.2 billion pounds of chemicals released, indicating the impact of manufacturing facilities.

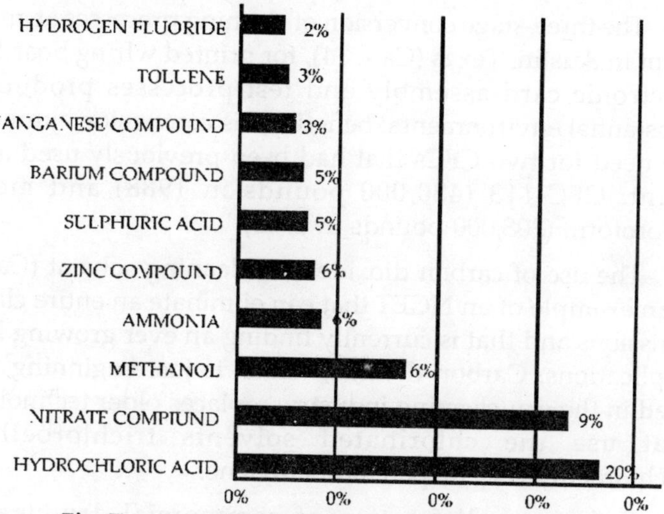

Fig: Top Ten Chemicals Released Excluding the Mining Industry (per cent of emissions).

NGETs have the potential to address many of the large-volume chemicals released by the chemical industry and other industries. Some technologies replace chlorinated solvents with more benign alternatives that could have an effect on the large amount of chlorofluorocarbons (CFCs) included in the TRI (for example, Cases 1, 14, and 18). Chlorine is one of the most released TRI chemicals, and the case studies having to do with the pulp and paper industry (Case 4m and 15) and water purification (Case 8) aim to reduce the use of this chemical. Some cases supply alternatives to many solvents, such as through the development of ionic liquids. However, replacing well-known solvents with these new chemicals is not yet fully understood.

Not all the chemicals in the TRI are exceptionally dangerous to the environment. Methanol, for example, has a relatively low toxicity, and none of the case studies addresses its removal. However, methanol's heightened persistence and widespread occurrence may be a cause for concern. The Twelve Principles of green chemistry call for reducing both the risk and the exposure to chemical substances whenever possible, regardless of the known risks.

The three-stage conversion of cleaning processes at an IBM plant in Austin, Texas (Case 14), for printed wiring board and electronic card assembly and test processes produced a substantial environmental benefit. This conversion eliminated the need for two CFCs that had been previously used at the plant: CFC-113 (430,000 pounds in 1988) and methyl chloroform (308,000 pounds in 1988).

The use of carbon dioxide as a cleaning solvent (Case 1) is an example of an NGET that can eliminate an entire class of emissions and that is currently finding an ever growing list of applications. Carbon dioxide, which is just beginning to be used in the dry-cleaning industry, replaces older technologies that use the chlorinated solvents trichloroethane, trichloroethylene, or perchloroethylene.

Only a small fraction of commercial dry-cleaning machines currently use the new technology, but such machines have the potential to terminate the use of chlorinated solvents at nearly 35,000 dry-cleaning establishments in the United States. This would eliminate nearly 51,000 tons of organic solvent emissions and 95,000 tons of other hazardous air pollutants every year.

Biochemie (a division of Novartis) developed a green chemistry process for an antibiotic intermediate, 7-amino-cephalosporanic acid (Case 4e). A reaction pathway involving two enzymes replaced a chemical process that used N,N-dimethylaniline, trimethylchlorosilane, phosphorous pentachloride, methylene chloride, and zinc salts. The bio-based process uses no toxic ingredients, is an aqueous room temperature process, and biological waste treatment can be used. Taxes on incinerated wastes from the former process provided an incentive for development of this NGET.

Significant environmental benefits come, as a number of cases indicate, when NGETs are developed for fine chemicals or pharmaceuticals. This is readily understood from the work of Roger A. Sheldon, chemistry professor at Delft University in the Netherlands. He developed the "E-factor" measuring the efficiency of chemical processes—the ratio of chemical waste to intended product. Sheldon found that commodity chemicals have an E-factor of between 1 and 5, while for fine chemicals the ratio is between 5 and greater than 50, and for pharmaceuticals 25 and greater than 100.

Health Benefits

NGETs provide health and safety benefits for workers in chemical and related production plants. Although the chemical industry has an excellent safety record, NGETs, to the extent that they reduce the need for toxic or otherwise dangerous materials during production, may reduce undue worker exposure. Many of the new processes, by operating at lower temperatures and pressures than conventional synthetic processes, can reduce further the dangers from potential accidents. The safe production of end-use chemicals provides one area of risk reduction.

A second is the production and manufacturing of new benign materials that pose fewer health and safety risks altogether. Many compounds produced today are known to cause problems to the users. For example, while the health and safety risks of producing the pesticides used in agriculture are relatively low, the application of pesticides poses an acute hazard to hundreds of thousands of farm workers, particularly in developing countries. The production of safer pesticides may reduce such problems.

NGETs also offer the workplace elimination of chlorine, chlorinated solvents, and ethylene oxide plus acrolein from the bio-based process for 1,3-propanediol, an intermediate for a new polyester. NGETs can also provide national security benefits. They can, for example, reduce U.S. dependence on critical materials (including oil and precious metals) from potentially unstable regions and reduce the hazards from chemical plants, storage facilities, and transport vehicles, eliminating potential targets for terrorists.

Shortly after the September 11, 2001 attacks, EPA reported that at least 123 U.S. chemical plants contained enough chemicals to result in a million casualties in nearby communities if attacked. Sudden changes took place in the way businesses perceived their exposures to chemical catastrophes. As an example, the Blue Plains Wastewater Treatment Plant swapped its 90-ton railcars filled with toxic chlorine for safer sodium hypochlorite to reduce the risk. Proposed legislation also created new incentives to reduce the use of hazardous chemicals. Over the past year there has been considerable interest among policymakers in reducing the risks chemical facilities might pose as targets for potential terrorist attacks.

Legislation proposed in the 107th Congress sought to prevent, control, and minimize the potential consequences of chemical releases by requiring firms to:

1. Identify hazards that may result from an accidental or criminal release;
2. Ensure safer design and maintenance of that source; and
3. Minimize the consequences of any such release. The legislation did not directly encourage the long-term phase-out of dangerous chemicals. However, the costs of meeting the required security levels associated with some chemicals might cause the chemical industry to consider the use of less-risky chemical alternatives.

Opponents of the legislation argued that additional oversight by the government could be costly and would not ensure safety. NGETs could provide a means to address security risks by removing the need to use and store some dangerous substances. Although Senate Bill 1602 was not enacted, such initiatives provide incentives for accelerating development of NGETs and related inherently safer processes.

One target chemical could be phosgene, a well-known chemical warfare agent that was used in World War I and is designated in several agency regulations as an acute hazardous substance. It can cause short-term irritation to the eyes, skin, and breathing passages. Repeated exposures can cause permanent lung damage. Accidents, spills, or terrorist attacks at sites using phosgene can affect surrounding areas. Reducing the use of phosgene can help reduce this security and occupational safety hazard. Alternative dimethyl carbonate (DMC) synthesis methods (Case 5) will displace large volumes of phosgene with materials such as methanol and urea, two relatively benign substitutes.

In some cases, process changes aimed at environmental improvement may have resulted in inherently less safe designs that introduce workers to greater health and safety risks. The Environment, Health and Safety Committee of the Royal Chemical Society points to the collection of vent discharge gases for incineration or for absorption, which resulted in explosions when the composition of the gases in the vent system entered the flammable range.

Ashford examined eight cleaner technologies in the International Cleaner Production Information Clearinghouse. He found that four worsened occupational safety and health, and four could have been further improved to provide greater safety and health benefits. One of the cases that worsened safety and health, "rapeseed oil extraction by enzymes," is related to the Biodiesel Case 4uu. The message is that process engineers must consider workers as part of the production process in designing NGET-based processes.

Economic Benefits

NGETs may provide economic benefits to firms and society at large by reducing the cost of meeting existing and future environmental and safety standards and security goals. In addition, firms may adopt NGETs in order to deploy new processes and offer new products that are economically beneficial to the firm in their own right, without consideration of their environmental, security, or health and safety benefits. This situation is particularly interesting because it suggests a potential for NGETs to play a broader role in environmental protection.

A good example is found in aluminum smelting. Energy accounts for approximately one-fourth of the costs to smelt aluminum. New inert anodes developed for aluminum smelting (Case 11) have enabled a reduction in the anode-cathode distance from 1.75 inches to 1 inch, in turn reducing the energy required for aluminum smelting by more than 25 percent. Potentially, this could translate into a U.S. energy savings of 6 trillion Btus annually by 2010, at a value of nearly $90 million per year.

Environmental benefits resulting from this NGET include the elimination of carbon and fluorocarbon emissions as well as organic matter generated during anode manufacture and consumption, and the reduction of cyanide formation and dust emissions. In addition, nonenergy cost savings amount to $20 million per year. Test cells are operating in Europe and North America.

A novel synthesis of 4-aminodiphenylamine (4-ADPA) used in the production of rubber chemicals can replace previous processes that involve the chlorination of benzene (Case 24). The new process eliminates more than 70 percent of organic wastes and more than 90 percent of inorganic wastes and wastewater. Although societal demands to eliminate chlorine were a factor in the development of 4-ADPA, the process yields significant cost savings to Monsanto and Flexsys, the joint venture of Monsanto (now Solutia) and Akzo Nobel, the firms that have commercialized it.

Monsanto, in producing the environmentally benign, nonselective herbicide Roundup® (Case 25) replaced the synthetic route to the key intermediate disodium iminodiacetate with a coppercatalyzed reaction that eliminated the use of hydrogen cyanide and formaldehyde. This reduced the waste flow to biotreatment by 800 million gallons per year while reducing manufacturing costs by more than 40 percent. Additional savings are still possible.

The alternative synthetic process for ibuprofen (Case 2) provides another commercially viable example. This process uses hydrofluoric acid as a catalyst and achieves a 77 percent atom economy, as opposed to the 40 percent achieved using the traditional manufacturing process. The greater throughput requires less capital expenditure and generates less waste.

This process was deployed by BASF in 1992 at one of the largest ibuprofen manufacturing facilities in the world, in Bishop, Texas, which produces 7 million pounds per year, nearly one-fourth of world's supply. The new process helped eliminate nearly 8.2 million pounds of waste per year; adopting it for all ibuprofen production, displacing the other five routes, could help eliminate 25 million pounds of waste each year.

Generation of Hazardous Substances

NGETs fully manifest themselves as a new approach to environmental protection when they provide environmental or other societal benefits at little or no cost. In such cases, firms may adopt the new technologies without the need for regulations or other mandatory measures. Ideally, government policy would then only focus on the R&D and information dissemination policies that promote the supply of NGETs. We have just seen cases where some environmental benefits are delivered with a net economic benefit.

Additional cases show instances where NGETs provide environmental benefits at minimal or no additional cost. Such situations are, of course, not unprecedented. The United States' 100-year decarbonization trend that has reduced the amount of CO_2 produced per unit GDP is largely a result of efficiency increases and fuel changes driven by economic forces. Nonetheless, NGETs offer the potential of extending such improvements to sectors where it has been lacking heretofore.

Many of the bioprocesses (Case 4) provide environmental benefits at little or no cost. Some products such as specific stereoisomers of vitamins, amino acids, fine chemicals, and pharmaceutical intermediates would be difficult or nearly impossible to synthesize by costly and inefficient traditional chemistry. Typical is the synthesis of Riboflavin or Vitamin B2 (Case 4a) which reduced gas emissions by 50 percent and water emissions by 66 percent while cutting operating costs in half.

Eastman and Genencor have developed a one-step fermentation process for the ascorbic acid (vitamin C) intermediate 2-ketogluconic acid (Case 4bb), replacing a fourstep conventional process and cutting plant investment in half. New biological processes helped a Chilean copper mine leach metals of value, such as copper, zinc, and cobalt, from a sulfide mineral (Case 4o). Bioleaching involves the use of bacteria, principally *Thiobacillus ferrooxidans, Leptospirillum ferrooxidans,* and certain thermophilic (high temperature) bacteria.

During oxidation, bioleaching places copper (or the metals of interest) in the solution phase; the oxidation residues are then handled to maximize recovery of the solution (within the volume and solution grade constraints of downstream processes), and the solid residue is discarded. Operators claim that bioleaching has environmental and economic benefits not available in more conventional processing technologies such as roasters, smelters, and pressure autoclaves.

Environmental benefits include the generation of environmentally stable residues and the elimination of noxious gases and toxic effluents. Economic benefits include quicker and less-onerous environmental permit and reporting processes and simpler and safer operations. Bioleaching also allows smaller projects to be developed more economically. Case 9 provides an example of NGETs that have helped eliminate some incompletely understood health and safety risks at an additional cost acceptable to the market. Until very recently, nearly all pressure-treated wood in the United States was preserved using chromated copper arsenate (CCA).

CCA contains two known human carcinogens: hexavalent chromium and arsenic. While EPA has not concluded at this time that CCAtreated wood poses an undue risk, it is widely believed that reduced exposures to arsenic, which could be dislodged or leached from the wood, is desirable. Treated wood is used in decks, playground equipment, picnic tables, boardwalks, fencing, and landscaping, putting children especially at risk. Public concerns about arsenic in drinking water, the desire of consumers for more environmentally benign products, and potential new regulations have all stimulated interest in an NGET that can eliminate the need for CCAtreated wood.

A new alternative wood preservative, alkaline copper quaternary (ACQ) provides the same level of protection as CCA against decay fungi and termites at a competitive cost. Unlike CCA, none of the components of ACQ (copper, dodecyl dimethyl ammonium chloride, or alkyl dimethyl benzyl ammonium chloride) are known to be carcinogens, and quaternary ammonium compounds have low mammalian toxicity.

Other environmental benefits include reducing the use of arsenic in the United States by 90 percent and eliminating the use of 64 million pounds of hexavalent chromium, as well as eliminating hazardous waste generation as designated by the Resource Conservation and Recovery Act (RCRA) along with the problems of disposing of CCAtreated wood.

Advanced oil and gas exploration technologies provide an example of advances in material technology that offer environmental benefits at little or no cost. Improved drill bits have cut drilling times in half and reduced the time drilling rigs must be on site, and thus their environmental impact.

Wide Variety of Reasons

Firms adopt NGETs because these technologies may help lower costs; however, a lower cost may be a necessary but not sufficient condition for adoption. In many of our case studies, there was some specific reason, often idiosyncratic, that the technology was adopted or not adopted. This suggests that there is no single set of government policies that might speed or encourage a broad range of NGET.

In particular, it is not always sufficient for an NGET to provide environmental benefits at no cost. Firms have many demands on their investment capital. They cannot invest in every option that saves them money. Rather, they seek the opportunities that promise the highest rate of return. The case studies here suggest that the existence of NGETs themselves is rarely the driver of a firm's decision to redesign its production processes. Firms instead may use available NGETs when they choose to upgrade or build new production facilities for other reasons. As a result, the adoption rates and timing of investments in NGETswill often be driven by factors external to the quality and performance of the NGETs themselves.

For example, the new process for making ibuprofen was not developed until the expiration of the Boots Company patent, which opened the market to generic competition. The loss of the patent spurred the firm (eventually to be BASF) to consider new, more-efficient manufacturing processes, including "greener" processes to fend off the new competition.

Companies, either the ones with patented drugs or those seeking to enter the market, will develop options as a drug comes off patent. One important option is to lower production costs. This is done by process improvements or by moving production to lower cost sites, often in developing countries. NGETs can in some cases offer a different route to lower costs, as was the case in the manufacture of ibuprofen. Although 20 percent of ibuprofen supply comes from Shansun Chemicals and Drugs Ltd. in India, the green chemistry–based process has allowed BASF to maintain a market share while remaining in Texas.

Likewise, the decision by Ube Industries to develop an alternative process for synthesizing DMC was a business decision to move downstream in the value chain and to take advantage of the corporate expertise in nitrite chemistry and the availability of ammonia, a raw material for DMC synthesis. Because this process bypasses the use of phosgene, Ube Industries has also gained recognition as a leader in dealing with environmental, worker health, and safety issues.

Should a market for DMC as a fuel additive develop, Ube Industries would also be able to capitalize on it through production at a recently constructed plant using its new synthetic process. Governmental regulations and other efforts to promote the demand for environmental protection have played a role in advancing many NGETs. Firms can seek NGETs to lower the cost of meeting actual or anticipated regulations. Regulations can also speed the adoption of low- or nocost NGETs because firms will invest in response to the timing of the government requirements.

For instance, regulations have stimulated investments in NGETs to avoid the use of banned substances, such as the use of chlorine in pulp bleaching, or to reduce the cost of compliance with environmental and worker health and safety regulations. The desire of firms to cut tax payments, including, for example, those in 15 states on perchloroethylene emissions, may also lead to further NGET innovation. Recycling programmes, take-back legislation, and recycled content standards also help to stimulate NGET markets. Some states have created incentives for NGET use; Nebraska and North Carolina, for example, provide incentives for dry cleaners to use the supercritical CO_2 process.

Taxes on incinerated wastes led Biochemie to develop a new route to an antibiotic intermediate. When use of chlorine was prohibited in the pulp and paper industry, various other oxidants were adopted, and when the storage of contaminated gypsum was prohibited in the Netherlands, Budel Zink adopted a bio-based zinc refining process.

On occasion, the cost savings offered by NGETs may drive firms to invest in the absence of other external factors. For instance, volatility in the world sugar market created an incentive to seek domestic replacements for sucrose. Significant work on recombinant thermostable enzymes led eventually to the $3 billion high-fructose corn syrup industry. Alternatively, an NGET may offer a firm the opportunity to enter new markets.

For instance, Cargill-Dow invested $750 million to develop a family of biodegradable polymers based on lactic acid to replace such materials as polystyrene and acrylics. DuPont licensed the DeSimone technology to use liquid carbon dioxide rather than CFCs or water as the polymerization solvent for producing a number of Teflon fluoropolymer products for reasons other than cost. Products with significantly improved properties were produced. Molecular weights and molecular weight distributions and endgroup chemistry could be more effectively controlled. This translates to an improved competitive market position.

Barriers are Significant

Economic models suggest that private firms often underinvest in the development of new technologies because they cannot capture the full benefits that accrue from such research. The knowledge they generate might be usefully copied or extended by other firms, benefiting society but not sufficiently helping the originating firm. In practice, firms will also fail to invest in some new technologies that would clearly save them money.

In many cases a firm may have existing capital stock that, while more expensive to operate, costs less overall than new technology primarily because of sunk capital. In addition, firms focus their scarce investment resources on technologies central to their core business strategy, often those expected to give them the highest rate of return. Thus, it is often insufficient for an NGET to merely save money for a firm; the technology must make the firm more money than some other competing investment.

Some NGETs promise significant benefits, but many have to date found only limited commercial applications, in niche or smaller markets. Societal benefits will be limited until they can be diffused more widely. Many studies have addressed barriers to the adoption of new environmental technologies.

For instance, the Environment, Health and Safety Committee of the Royal Society of Chemistry provided one recent summary of the barriers to developing green chemistry:

1. The lack of global harmonization on environmental regulations;
2. Rigid notification and authorization processes hindering new product and process development;
3. Need for speed and certainty of results caused by short-term planning;
4. Economics—the green chemistry approach may be more costly, though this is not invariably the case;
5. Accounting practices which do not consider all costs;
6. Difficulty in getting research funding;
7. Insufficient guidance on best practice for green chemistry;

8. The low profile of more sustainable chemistry in schools and universities; and
9. A culture geared to looking at the product itself rather than the overall process and life cycle. We observed many of these barriers and others in the case studies.

Four types of barriers were especially evident in the cases:

Need for additional research, technology development, or process engineering

In nearly all instances when a technology is initially developed and then commercialized, additional research or technology improvements will be needed to gain or maintain competitive advantage. Two competing groups, Cool Clean Technologies and Global Technologies, are introducing dry-cleaning systems using liquid carbon dioxide (Case 1), and each seeks to improve the economics and the performance of the washing systems, especially the surfactants.

DuPont is piloting a bio-based process for 1,3-propanediol (Case 4cc). While doing this, the firm is using a chemical-based process to supply enough diol to react with terephthalic acid and commercialize Sorona® polypropylene terephthalate [PPT, or poly(3-GT)]. Both chemical engineering and metabolic engineering efforts are required to continue the scale-up of the biobased process. In some situations a significant barrier may be in the gaining of resources or funding to do such additional research or development.

There will be instances when research on an NGET conducted in a university or government lab is not continued by or transferred to an industrial laboratory. The firm may not want to allocate R&D funds to the development; it may feel that even though the technology is promising, the risk is too great versus other R&D targets; or it may feel the technology is still untested. Case 4w, the bio-based process for adipic acid is one example. In some situations, promising research on NGETs fails to receive federal funding.

Some argue that priorities for environmental research often fall below those for research in the life sciences or other research funded by numerous agencies through congressional appropriation committees. There is no coordinated effort, for example, to eliminate the top volume pollutants; the emissions come from many different firms, each with its own control mechanisms, related economics, and incentives for research, with no coordinated federal effort to tackle such major common problems.

Need to Surmount Infrastructure and Integration Barriers

Solutia (formerly Monsanto) plans to use by-product nitrous oxide from the adipic acid process to manufacture phenol (Case 21), but matching the output of the two processes becomes a challenge. When quantities of nitrous oxide are limiting, other routes to phenol or other sources of nitrous oxide must be found; this could greatly affect the economics of the phenol process. Startup of this process has been delayed, indicating the extent of these barriers.

Systemic barriers have also limited NGET diffusion for the depolymerization of recycled plastics, particularly polyethylene terephthalate (PET) and nylon 6 (Case 3). About 750 million pounds of PET are used in the United States each year, particularly for soft-drink bottles and other food containers. Nylon is prevalent in carpeting and is increasingly used in automobiles, but its use results in large amounts of currently nonrecyclable carpeting "fluff." One key hurdle to plastics recycling is the difficulty in separating the different components of the waste stream so that they can compete with virgin materials. At present, only a small fraction of plastics is returned to highgrade uses.

The rest is used for low-grade applications and does not replace much raw material. The lower cost of virgin materials in particular can limit NGETs for plastics recycling. A joint venture of DSM Chemicals and Honeywell International to depolymerize recycled carpeting and produce 100 million pounds of caprolactam each year from 200 million pounds of recycled carpet, shut down after two years of operation when plummeting prices for virgin nylon made operation of the plant uneconomical.

This failure indicates some of the problems NGETs often face when moving beyond recycling markets or niche markets in general. It is difficult to compete in commodity markets dominated by large existing producers where prices can fluctuate strongly. NGET plastics recycling also relies on a less certain source of supply than that of its competitors using virgin material, who typically buy from a small number of large petroleum firms. In boom times, it is straightforward to quickly garner the supplies needed to expand production.

In contrast, plants such as Evergreen (for nylon 6 recycle) rely on a diffuse network of households and commercial office buildings willing to deliver their old carpets. While Evergreen suffers along with its larger competitors during periods of low prices, it cannot as easily scale up its needed raw materials when boom times offer large profits to increased production.

Need to Make the up-front Investment

The use of CO_2 as a solvent in the dry-cleaning industry also requires higher up-front capital costs and specially trained operators. While the operating costs of the new technology may be lower and the payback periods are similar, the CO_2 machines are twice as expensive as the traditional machines that use perchloroethylene. These significant up-front costs pose a barrier to its use. Working with supercritical carbon dioxide necessitates lower temperatures and higher pressures that go beyond the capacity of normal dry-cleaning machines. Special care is also needed to prevent the release during cleaning of large quantities of CO_2, which can inhibit breathing by machinery operators.

Regulatory Barriers

While regulations can stimulate NGETs, the design and implementation of regulations can often hinder implementation of NGET-based processes. When strict timetables are required to meet new emission limits, there is little incentive to develop a greener process. DuPont manufactures phosgene at its Chambers Works plant in New Jersey. The New Jersey EPA placed strict requirements on carbon tetrachloride emissions which would require a $2 million incinerator. Negotiations led to an extension of two years. In that time, work on new catalysts reduced emissions to less than the required level while promising to reduce the emissions ultimately to zero. Such regulatory flexibility is not always forthcoming.

The way in which patents are enforced greatly impacts the incentive to develop new routes to an "active pharmaceutical ingredient" (API). As we saw in Case 2, the incentive for developing the new route to ibuprofen was the expiration of the fundamental product patent. In parts of Europe, companies cannot develop generic versions of brand-name products that still have patent protection. In the United States, however, because of the Hatch-Waxman Act, manufacturers of an API can develop a new route before the patent has expired.

Thirty-five blockbuster drugs are coming off patent between now and 2007. The combined sales of $82 billion for these 35 can create incentives for process research. For example, the patent protection for Zocor, Merck's cholesterol lowering drug (annual sales of nearly $7 billion), expires in 2005. Hovine in Portugal has teamed with a Korean firm to develop a fermentation process to supply the API to displace Merck.

LONGER-TERM OPPORTUNITIES FOR NGETS

In addition to the important but limited near-term benefits noted above, NGETs have the potential to provide broad and significant benefits over the coming decades, through further development, greater diffusion, and the introduction of new applications. It is of course difficult to estimate with accuracy the long-term benefits offered by advances in technology, particularly when effective application involves adoption by firms in competitive markets. When policymakers consider the role new technologies can play in helping to achieve long-term goals, it is useful to present a range of possibilities. In this report we provide indications of these possibilities.

Over the long-term, NGETs and green chemistry offer many promises, including substantial reductions in the environmental footprint of many chemical processes, improvements in the health and safety of those exposed to chemicals, and enhanced security at facilities with hazardous materials. These and other benefits of NGETs will arise from a combination of technology innovation and investments in new processes. The case studies suggest a range of paths that environmentally friendly chemistries will follow as they make their way into the marketplace.

Some paths emphasize market pull. Demand for new environmental processes, driven either by regulation or societal pressure, will spur the development and deployment of NGETs. Demand pull need not generate new research; often well-known technology can be used in new applications. In paths emphasizing technology push, new research or new applications demonstrated in one area may suggest other applications where a potential need was previously unrecognized or seemed impossible to address effectively. The combination of demand pull and technology push will create the long-term opportunities offered by NGETs.

New Processes and New Chemicals

In many cases, known and sometimes long-standing environmental and other problems can be addressed by new science and technology or novel applications of known technology. Our case studies show a wide variety of NGETs developed in response to new regulations or other pressures signaling a societal demand for new levels of environmental performance. Many of these have significant longer-term possibilities.

As has been noted, the chemicals and allied products industry is responsible for more than 600 million pounds of waste included in the TRI compiled by EPA. Many of these chemicals are manufactured with processes that utilize older synthetic pathways fine-tuned over many years without significant consideration given to environmental impacts. Our case studies demonstrate that in some cases NGETs, often applying existing science, can prove more cost effective than even these older highly optimized processes. The new synthetic pathway used by BHC to manufacture ibuprofen provides the most well-known example.

In the future, an NGET that eliminated phosgene from the creation of DMC could have broad benefits. By reducing the environmental as well as the economic costs of phosgene in the synthesis, such an NGET might enable DMC's use in a variety of synthetic processes as a methylating agent, replacing less environmentally friendly reagents such as methyl chloride and dimethylsulfate.

The potential substitution of DMC for phosgene in the synthesis of polyurethanes, polyamides, and polycarbonates, processes that account for more than 90 percent of the demand for phosgene, could result in an expanded market for DMC, thus lowering its cost while significantly reducing the demand for phosgene with its attendant hazards. In this case, environmental benefits occur both when a new nontoxic biodegradable chemical (DMC) is produced and when it is used in the synthesis of other products.

The fuel additive market provides another example of the potentially dramatic benefits of substituting greener chemicals for others currently in common use. When used as a fuel additive, DMC has a higher oxygen content than methyl tertiary-butyl ether and is less toxic and biodegradable. However, at present, DMC suffers from a prohibitively higher cost. If technology can reduce these costs through a new greener synthetic procedure, DMC could reduce CO, NOx, VOCs, and particulates beyond the levels with existing fuel additives.

NGETs might also provide broad, long-term benefits by enabling use of new environmentally friendly refrigerants in response to the push to rid the environment of CFCs. Case 18 suggests that newcompounds are available to reduce energy use substantially through more efficient, environmentally friendly refrigerants. Such compounds could potentially remove the CFCs and recently developed CFC alternatives currently used in refrigeration while at the same time increasing the efficiency of refrigeration and thus reducing the use of energy. Reductions in energy use for refrigeration obviously constitute an additional benefit to the environment.

Other NGETs based on greener paths to hydrogen peroxide and hydrogen peroxide activation could potentially replace the use of chlorine in the pulp and paper industry. This industry currently bleaches pulp with a five-step process alternating with different oxidants. H_2O_2 has the potential to replace some of the steps in the process that use Cl_2. But H_2O_2 alone cannot replace ClO_2 in the final steps. Case 15a suggests that the activation of H_2O_2 with the use of a chelated iron compound can make the use of hydrogen peroxide more efficient and competitive with the ClO_2 -based process.

The societal interest in the removal of chlorine from the bleaching process has spurred new interest in using small molecule activation by inorganic catalysts. If successful, these NGETs could remove the chlorine from circulation and thus the toxic chemicals resulting from its use. To capture many of these long-term benefits, new inorganic catalysts need to be tailored to the specific hydrogen peroxide applications.

A better and cheaper synthesis of H_2O_2 (Case 6) could have important spillover effects for a host of industries. Such greener oxidants would benefit the pulp and paper industry as well as the chemicals industry, where demand has grown steadily over the decade. For example, the demand for H_2O_2 as an oxidant in printed circuit board etching and semiconductor manufacturing and in wastewater treatment has increased substantially in recent years.

Water purification (Case 8) highlights the need and potential for displacing the chlorine currently used in 98 percent of all water treatment plants. More than 13 million pounds of chlorine are currently used in water purification, posing not only health risks but security risks as well. Replacing current processes in competitive markets can of course be difficult even with NGETs based on well-understood science. For instance, the chemistry for returning polymers to monomers has been known for quite some time. In principle, this technology could significantly enhance the level of plastics recycling (Case 3) and could yield broad social benefits.

The rubber and plastic industry accounts for nearly 100 million pounds of waste, of which more than 99 percent is emitted to the atmosphere. Increased recycling could reduce the amount of virgin materials used by the chemicals industry to supply the rubber industry. Landfills could also be reduced.

Carpets for example, account for nearly 1 percent of all landfill, or more than 4 billion pounds of nonbiodegradable annual waste consisting largely of potentially recyclable polymers.

However, NGETs for returning this waste to monomers that can serve as raw materials to the rubber and plastics industry have been unable to compete in the marketplace because of the large price swings and significant investment needs of commodity markets for raw materials. In particular, firms can find virgin raw materials a preferred source of supply because of ready availability even in times of increased demand.

Opening New Green Pathways

In addition to situations where known societal problems spur the application of NGETs, there are also many cases where new green chemistries offer the promise of preventing potential environmental problems that have not yet been identified or risen to the level of a serious problem. Research in significant areas such as catalysis and biotechnology can suggest unexpected environmental, health and safety, and security benefits.

The use of supercritical carbon dioxide as solvent, genetically modified enzymes as effective biocatalysts, or nucleophilic aromatic substitution for hydrogen (NASH) all provide examples. The long-term benefits of such new solvents, reagents, and reaction conditions may provide substantial environmental, security, and health and safety benefits and avoid a range of heretofore unrecognized problems. In many of these cases, significant research remains to be done before these benefits can be fully realized.

As one important example, organic solvents are used across many industries for precision cleaning, medical device fabrication, and the dry cleaning of garments. These industries use more than 30 billion pounds of halogenated and other organic solvents per year. Some of the environmental impacts of solvents are well known. Solvent use in general is a leader in the TRI, and solvents comprise eight of the top ten TRI chemicals that are sent off site for remediation. The worldwide solvent market has been in decline for some time because of the restrictions on many of the more popular VOCs and chlorinated products.

While the markets for oxygenated solvents have seen recent expansion, especially in the Far East, other areas of solvent use have been curtailed. Regulatory pressures on VOCs in general, and specifically on those used in paints and coatings, will constrain these solvent markets. Nonetheless, not all potential impacts of solvent use are known with confidence. For instance, some of the solvents in the TRI list are not necessarily toxic nor do they pose serious immediate environmental risks. However, because such large volumes of these chemicals are used across industries, the effects of the combined releases are not completely understood.

The general principles of green chemistry call not only for removing the toxic and dangerous compounds from syntheses but also for reducing the total amount of compounds released into the environment regardless of the toxicity, as a hedge against unknown effects. To that end, several new technologies offer promising alternatives, including the aqueous biphasic systems and room temperature ionic liquids highlighted in Cases 16 and 17. Supercritical carbon dioxide has the potential to develop into a useful solvent to replace some of the TRI chemicals.

Carbon dioxide has unique characteristics as a solvent, but a variety of technical barriers must first be overcome for it to be used ever more broadly. For example, carbon dioxide has a wide range of potential applications, but each application requires a good understanding of the ideal pressures and temperatures that can be used for specific solute. Firms must invest in research to determine the optimal conditions for each ase. Carbon dioxide will become more pervasive as each new cess encourages other firms to explore the technology.

The movement of synthesis toward better control of molecules through catalysis is a common theme among f our case studies and is an area of technology development with the potential for substantial long-term environmental implications. Both the safer production processes and potential markets of DMC and H_2O_2 provide examples. For instance, new iron catalysts will activate the small molecules of H_2O_2 to provide the oxidation potential needed to bleach wood pulp. Without the catalysts, the hydrogen peroxide may not be able to reach the same levels of bleaching power as conventional chlorine-based bleaching.

Many of the iron catalysts under development as activators mimic the type of biological processes found in nature, where natural enzymes activate hydrogen peroxide for use in oxidation reactions. Developing technologies to exploit natural processes may provide far-reaching environmental implications. The development of better synthetic pathways has increasingly exploited the spillover of biological concepts into chemical science and engineering. The bioprocesses highlighted in this study were developed with the hope of moving closer to what biological systems do well—manipulating a relatively small number of starting materials (e.g., O_2, N, CO_2) and transforming them into a multitude of complex compounds.

Many of the toxicity, health and safety issues, and environmental problems associated with current synthetic pathways stem from the complex starting materials and intermediate compounds produced throughout a synthetic pathway. Starting with simple and easily found materials, such as those already present in the atmosphere—N, O_2, CO_2 —and manipulating them in single steps into complex compounds may enable the atom economy necessary to decouple society's chemical processes from environmental impacts. Moving from the complex starting materials and simple reactions to simple starting materials and complex, highly specific reactions may afford the renewable feedstocks and reduced footprint necessary to ensure that the environmental, security, and health and safety risks can be reduced.

Bioprocesses highlighted in this report have sh promise as greener synthetic pathways. The benefits ster the complex reactions that biological systems are eff performing compared with conventional chemical technologies. Bioprocesses reduce energy use, quantities of feedstocks, and the amount of waste generated compared with alternative nonbiologically driven processes.

Regulatory pressures have promoted the use of biobleaching and biopulping in the pulp and paper industry for environmental reasons, and advances in biocatalysis (especially in enzyme development) have significantly increased yields and the economics of many processes. Bioprocesses have also enabled the manufacture of many chemicals and pharmaceuticals previously very difficult to produce while leading to reduced waste and improved economics.

Enzyme developer Maxygen believes that $50 billion of the $800 billion global commodity, specialty, and fine chemical markets are readily addressable by bioprocesses. It also says that another $200 billion has been identified as potentially addressable by biological approaches in the next 10 to 20 years. Much of the research performed to date on bioprocesses is now more readily distinguished as part of the cadre of NGET tools. The reduced energy requirements, specificity of reactions, and benign reagents typically involved may address the inherent problems with traditional synthetic pathways. This section presents conclusions based on our case study results and observations based on the analyses of barriers.

NGETs have the potential to make a significant contribution to a new approach to environmental protection. NGETs represent an important class of pollution prevention technology that redesigns products and processes at the molecular level to reduce or eliminate the use of hazardous materials. Our case studies demonstrate that NGETs have already provided a wide range of environmental, security, health and safety, and economic benefits. In some cases, NGETs can provide these benefits for no additional cost or even at a lower cost compared with today's practices. And because NGETs are relatively new technologies, they may help policymakers ensure environmental quality with less reliance on costly regulations as they mature.

The nation's economy increasingly relies on a wide variety of chemical products and processes. Progress in slowing the use of potentially hazardous substances has not kept pace with other positive environmental trends over the past 30 years. Finding new ways to ensure that needed chemicals pose no or little risk requires new thinking into the means by which they are produced and employed. Over the past decade, green chemistry, an important source of NGETs, has begun to provide alternatives to chemicals that are known to be dangerous or to potentially pose risks.

The case studies suggest that green chemistry technologies also have potential to expand beyond their current niches and provide significant long-term benefits spanning the four areas highlighted in this study. NGETs may help eliminate emissions from broad classes of chemicals through inventive solvent choices; toxic reagents may be nearly completely displaced by an environmentally benign alternative; the security risks of dangerous chemicals may be avoided through the creation and adoption of less hazardous alternatives; and synthetic processes may be more readily adapted to take into consideration environmental concerns.

In addition to addressing currently known risks, NGETs can also help address those that are imperfectly understood. The principles of green chemistry guide scientists and engineers toward processes that intrinsically embody more of the process inputs in the final product and thus are more efficient and generate less waste. Such principles can eliminate potential environmental problems before they arise.

Firms Adopt NGETs

Driven by technology push alone, however, NGETs are unlikely to reach their full potential. The case studies demonstrate that firms adopt NGETs for many reasons, ranging from business decisions to pursue new production processes or product lines to the need to respond to regulatory pressures.

Firms adopt NGETs because they are the most cost-effective means tomeet some particular end. In each firm, however, NGETs must compete for resources with a host of other investment opportunities. To garner scarce investment dollars, it is rarely sufficient for a new technology to merely save the firm money. It must be among the options that move the firm most aggressively toward its corporate goals.

Policymakers

This report made no attempt to assess the need for or the types of policies the federal government might implement in response to the potential offered by NGETs and the barriers to their implementation. Nonetheless, the report suggests several observations about potential policy actions.

A number of different types of government policies can help NGETs reach their full potential. One such policy area is that of information dissemination. Our case studies suggest that there are important spillovers across and among different areas of green chemistry. Applications of broad processes such as biocatalysis in bioprocessing in one area may suggest applications in another. When a firm develops some new product or process, it does not generally scan the full range of possibilities but rather tends to scan a more narrow range close to current practice. In part, research and design is expensive, so innovators will focus the majority of their efforts on those design choices where they expect to find promising answers.

Publicizing the scientific principles and successful applications of green chemistry may cause firms to guide their internal research toward innovations that become NGETs. Examples of such information dissemination policies include partnerships involving industry to develop environmentally benign processes, such as those with the Department of Energy's (DOE's) Office of Industrial Technologies; awards programmes such as EPA's Presidential Green Chemistry Challenge Awards; fellowships and other funding for green chemistry at the nation's universities; and encouraging the work of professional, scientific, and engineering societies in education and training related to green chemistry.

The government also has an important role to play in supporting basic research. New areas of chemistry and other sciences, still in their relative infancy, may lead to significant changes in a vast number of processes across the economy. Basic research may help to unlock their potential. At present the federal government funds a range of research potentially relevant to NGETs such as the EPA/NSF (National Science Foundation) Technology for a Sustainable Environment programme. Additional relevant programmes have been proposed.

The National Coalition for Advanced Manufacturing has called for a three-year, $5 billion basic manufacturing science and technology initiative, which includes research on manufacturing technologies for energy efficiency and environmental quality. We have not made an attempt to survey what the government is funding relevant to NGETs and then to assess this funding against the most promising avenues and areas of greatest need. Such a complete survey of federal funding is difficult to do given the diverse funding sources, the variety of programmes and objectives, and the uncertainty that energy, life sciences, or chemical engineering programmes will result in green chemistry or NGET advances.

Finally, the widespread diffusion of NGETs may require policymakers to also address demand pull by providing appropriate incentives for the adoption of NGETs. Crafting such incentives will be an important challenge. On one hand, NGETs present policymakers with questions common to environmental policy. For instance, NGETs may diffuse relatively slowly if internal cost savings are the only incentive the firms have to adopt them. Policymakers will need to determine whether the potential benefits of NGETs warrant other incentives to speed their adoption.

However, there are opportunity costs to encouraging firms to allocate their capital and technical personnel to develop NGETs. Resources spent on such pursuits are not available to develop other products and processes. Policymakers will need therefore to gauge what level and type of effort seems appropriate to further the development of NGETs. Should the approach be technology push or demand pull, fund research, increase the extent of regulation, or some combination?

There are also some areas where NGETs pose particular challenges. These include the focus on process technology, the uncertainty of future process applications, and the extent to which technology spillovers broaden the adoption of new manufacturing processes. For example, the focus on process technology is often invisible to consumers and thus relatively decoupled from any emerging market demand for cleaner products. Many government policies also focus on industrial products rather than processes that produce them. For instance, the EPA premanufacture notices (PMN) required of chemical manufacturers focus on the potential toxicity and hazards of new products rather than the greenness of the process by which they will be made.

NGETs also present policymakers with a problem of information asymmetry. There are a vast number of products and processes for which the adoption of NGETs might provide important environmental, security, safety, and economic benefits. Only firms have the information that can allow them to discover which products and processes offer the most gains from NGETs. But society as a whole may reap a large fraction of the gains from such applications. Some of these benefits come through reducing risks that are presently imperfectly understood or that seem too costly to control.

Clearly, policymakers should not threaten to regulate in every area where NGETs might provide some benefits, but many potential benefits may remain undiscovered without some expression of societal demand. Thus, policymakers must devise means to appropriately enable and encourage chemists, chemical engineers, and other potential innovators to explore the effective new scientific tools at their disposal and bring them to bear most effectively on the array of applications for NGETs.

2

Biodiversity Value and Loss

BIODIVERSITY RESOURCES

Biodiversity is the diversity of different species together with genetic variation within each species in a given area. Myers called it the 'primary source', and it is material vital for sustainable development. It is vital to improve biodiversity conservation because:

- Crops, livestock and pharmaceutical products are constantly challenged by a changing environment and evolving pests and diseases, and must satisfy new demands and fashions; biodiversity is needed to meet these challenges.
- Without the 'raw' genetic material from plants and organisms of all sorts there will be great difficulties developing new crops, pharmaceutical products (drugs and antibiotics), new fish suitable for aquaculture, improved and novel livestock, bacteria and yeasts for fermentation or biotreatment of waste or composting, and many other innovations.
- Biodiversity has philosophical and aesthetic value and may inspire new ideas and scientific advances.
- A case can be made that humans have a moral/ethical obligation to protect biodiversity.

- Sustainable development dictates that humans should pass on to successive generations at least the same amount of biodiversity riches they have enjoyed.
- Biodiversity-poor environments may be less resilient and less able to recover if disturbed. However, there is considerable debate on this issue.

Human knowledge is full of gaps; we are not sure what may be needed to ensure a stable, secure and sustainable future environment and human well-being. Biodiversity cannot be recreated once lost; it is a rich resource which should be held in trust for the future. For some time biodiversity losses have exceeded natural rates, and it is getting worse - to the extent that many argue that the world's greatest mass-extinction is under way and that humans are responsible.

Modern routes to improved agricultural productivity have mainly been via reduced biodiversity: the replacement of natural vegetation with areas dominated by very few crop species. Human activities are also reducing the diversity of insects, weeds and other organisms. Large agro-industrial corporations promote fewer crop varieties, and these are often bred in such a way or genetically engineered so that there is no chance of farmers saving viable seed. Growers become more dependent on business, and it is causing a reduction in food security.

For most of the time that humans have practised agriculture, domesticated varieties have grown in fields with wild ancestor species nearby, so that there has been ongoing cross-breeding which generates useful new varieties and maintains genetic diversity in the crops. Today farmers are compelled by commercial forces and legal controls to plant a limited number of varieties, so older types are lost while environmental degradation depletes the surrounding pool of wild species.

Agricultural seed and biotechnology companies are effectively asset-stripping biodiversity from the wild and, in some cases, from the seed banks. There has been debate about the morality of this conversion of public resources to private. Morals aside, it is clear that companies that develop new crops and biotechnology seek to recoup their investment in research by patents and similar measures aimed at protecting intellectual property rights (IPRs) and by restricting access to their gene banks. These companies are mainly large corporations and their source of genetic material is often developing countries.

Many of the varieties of food crops in the USA have already been contaminated by DNA from genetically modified organisms (GMOs). There is a risk that this could take place worldwide. It could also affect material in gene banks because viability declines and stored material is regrown and seed re-harvested at intervals; during regrowth seeds could be contaminated by pollen on the wind or carried by insects. In 2003 the Cartagena Protocol on Biosafety came into force; a legally binding international agreement which governs the transboundary movement of living GMOs accessed March 2004. Hopefully, this will help counter GMO threats to biodiversity.

What is needed is a reliable open-access system of biodiversity conservation, an improved version of that which has been maintained by the international institutes funded by the World Bank, charities and some governments, which led the Green Revolution. Unfortunately, the latter are suffering shortage of funding, and even important botanic gardens like Kew struggle for adequate resources. What is needed is international agreement to create a permanent, legally binding biodiversity conservation system, which involves all countries.

Failure to do so could result either in increased plunder of developing countries by powerful corporations seeking excessive control of genetic resources, or each country seeking strict sovereign rights over its biodiversity, which it can then sell to the highest bidder. Recently there has been lobbying to provide some rights for the source areas and peoples who have traditionally used the genetic material.

The Convention on Biological Diversity was negotiated just before the 1992 Rio Earth Summit, and came into force in

1993. It accepts that those countries with rich biodiversity would conserve their resources better if they could make money from it. As well as encouraging conservation the Convention supports sustainable use of biodiversity and equitable sharing of benefits. In some cases trade in valuable species might be used to pay for conservation; the terms of the Convention on Trade in Endangered Species need reforming to achieve such ends, and to improve its effectiveness.

In spite of sluggish support from the USA, Canada, Australia, India and New Zealand, there have been efforts to agree that in return for open access to biodiversity, centres would ensure their collections were freely available. Often the progenitors of domesticated species of plants, livestock, fish and so on have restricted distributions in the wild, and can easily become extinct as land is developed or rivers and water bodies polluted or drained. For example, a large proportion of the world's total cocoa diversity is found in a relatively limited area of the upper Amazon where it is vulnerable to oil exploitation, logging, squatter settlement and forest fires. Roughly 40 per cent of the world's remaining tropical rainforest lies in the Amazon Basin; in the last few years Brazil has invested large sums in highway paving and new infrastructure projects, notably the Avança Brasil programme. These and the expansion of crops like soya are likely to accelerate biodiversity loss.

How can remaining biodiversity be best protected? The main ways are to promote conservation, and to discourage property rights, trade practices, fashions and investment which act to destroy diversity. In addition, efforts should be made to promote awareness of environmental changes which might affect conservation areas and to counter any pollution which threatens biodiversity.

People are subjective in selecting what to protect. An organism may be saved because it is perceived to have potential economic value, or it seems deserving, or conservation occurs because a site or the biota are sacred. As

Tisdell noted, the loss of the dodo is often mourned, but the loss of numerous other species less well known or spectacular has had minimal public attention. Those seeking funds for conservation may find it less easy for species with less clear economic value or lacking attractiveness to the public. Where biodiversity is clearly seen to have the potential to generate future profits, commercial interests may undertake conservation. For other species, various forms of aid will play a crucial role in conservation.

Reasonable stocktaking of biodiversity resources is an important first step. If possible, an assessment should then be made to determine what is vulnerable. Next, efforts should be made to identify the root causes of loss; these are often indirect, cumulative and complex, and so difficult to trace. Once all that has been done it should be possible to develop a biodiversity management strategy.

The fight to conserve biodiversity has intensified since the mid 1980s, and there are now many specialist NGOs which focus on specific ecosystems or species, as well as those with more general biodiversity conservation interests. A body active in seeking to conserve tropical rainforests is the World Rainforest Movement, an international coalition of NGOs that has been active since 1987. Biodiversity conservation was given centre stage when the World Bank launched a Biodiversity Action at the Rio Earth Summit in 1992. But much more still remains to be done.

Often conservation and livelihoods appear to conflict: efforts to protect forests may be greeted by locals involved in the timber industry as a threat to their livelihoods; attempts to restrict fishing methods or access to areas that have been traditionally used can result in opposition. Where conservation areas are seen to harbour pests or diseases that affect surrounding areas people may object and even trap animals. Large animals like elephants, large cats and those that migrate cause special problems because of their mobility and because some are a threat when they make forays from reserves into surrounding areas. Marine mammals and migratory fish also pose conservation problems because they need global protection.

Many conservation authorities argue that there is a need to involve local people and gain their support, to ensure they are protected from nuisance or dangerous species and, wherever possible, to offer them livelihood opportunities associated with conservation. If not involved locals may feel alienated and then poach or resist conservation in other ways. Local people are increasingly involved in policing, administering and servicing conservation areas; they can bring to bear considerable traditional knowledge and are adapted to local conditions.

There are, however, some very strong critics of the present fashion to integrate conservation with economic development and the encouragement of local participation. In West Africa such approaches may not be as successful as proponents claim, and Oates argued that conservation should be for the intrinsic value of the biodiversity, and not part of development. Clearly, participatory conservation initiatives need to be carefully planned and monitored, and not simply established according to current development or aid agency fashions and then left unsupervised.

Recently, a Brazilian ethnobotanist and agriculturalist working in eastern Amazonia produced a cartoon picture book, 'Fruits and Plants used for Life', aimed at influencing local people to value biodiversity. This was a bold move for academics to invest precious research into a publication, which might offer little professional kudos, even though it is of great practical value - a dilemma worldwide. The cartoons illustrate to farmers with little education what return they would get if they retained trees, rather than sold them to loggers.

The longer-term benefits are clearly contrasted with immediate one-off gain. This blending of scientific study and local knowledge, presented in a very accessible form, promises to improve livelihoods and biodiversity protection - and might well be a strategy which could be profitably applied elsewhere in developing countries for, say, soil conservation and other 'accessible' environmental management advice. As mentioned earlier, there are often attempts to combine biodiversity conservation with sustainable livelihoods, for example:

- *Tolerant forest management*: the extraction of products and some cropping, whilst striving to maintain as much of the original forest or other vegetation cover as possible.
- *Extractive reserves*: conservation areas where local people or other approved groups can remove products in ways that do minimal damage. The extraction may help pay for the conservation. Brazil first created extractive reserves in 1990.
- *Green tourism*: efforts are made to reduce environmental impacts and possibly use some of the profits for environmental management; the tourism is often dependent on natural features or wildlife. Green tourism includes: golf courses, which restrict pesticides and herbicides and seek to encourage wildlife; waterspout resorts, which try to protect reefs and marine life; and archaeological sites, which pay for roads and generate funds for local communities.
- *Ecotourism*: this is a stronger form of green tourism, which contributes a significant portion of profits to environmental management.

Typically local people act as guides, and staff the accommodation and other services; the attraction is usually wildlife, scenic beauty, archaeology or tightly controlled hunting and fishing or photographic tours. An attempt is made to keep the impact of tourists and support facilities to a minimum, and educate the tourist to respect nature. Sometimes, tourists even pay to work on environmental projects.

Community-based conservation has been spreading since the mid 1990s, with a number of aid agencies supporting initiatives; for example, the Communal Areas Management Programme for Indigenous Reserves in Zimbabwe. Typically efforts are made to involve and empower locals, tap their knowledge and skills, and ensure institution building provides stable foundations for supporting conservation.

Frequently there are excessive claims for the benefits of poorly managed ecotourism. There may also be a problem of seasonality, so that ecotourism only provides a satisfactory livelihood for part of the year. As with other forms of tourism, there is a certain amount of vulnerability to sudden changes in fashion or scares about travel caused by distant crises, which can suddenly hit profits and any environmental management which has come to depend on them.

Conservation can also be linked to other benefits; for example, a forest reserve may be managed to lock up atmospheric carbon to counter global warming or protect a catchment providing drinking water or flows to irrigation or hydroelectric plants. In steep areas vegetation cover is vital to prevent mass movement onto land below. There is also a crucial link with commercial research and development. A conservation project might even be integrated with several beneficial activities; an example in a developed country is the Great Barrier Reef World Heritage Area Strategic Development Project, which links conservation with over 60 organisations in reef-related food production, job creation, recreation, cultural heritage protection and so on.

Developed countries benefit from funding biodiversity conservation research and monitoring. Where there are commercial pressures leading to biodiversity losses, developed countries can support efforts to seek alternatives: for example, developing substitute products. Developed countries can fund advertising to counter fashions that endanger species. Treaties and controls can help reduce the trade in endangered species. Sales controls and taxation can reduce trade in endangered species and fund conservation activities, as can tourism taxes, airport taxes and levies on traded items - provided they are not judged to be a trade barrier. The processes of globalisation and free trade agreements need to be closely monitored to ensure they do not trigger biodiversity losses.

Nowadays, people are using more products derived from areas well away from where they live, so citizens in developed countries have an impact on poor nations as a result of new trends like buying exotic fruit, vegetables and flowers. They should be made more aware of this and of their obligations to pay something toward resolving any problems their consumer habits cause. Structural adjustment packages of financial assistance for countries in economic trouble have been agreed on the condition they impose financial austerity measures.

Bodies like the International Monetary Fund and World Bank have promoted structural adjustment since the 1980s, but the conditions they impose have had negative impacts on the environment of some countries. For example, farmers may have lost subsidies and turned to more exploitative land use off-farm, such as charcoal production, which damages the environment and wildlife. Another problem for biodiversity in developed and developing countries is the replacement of large swathes of natural vegetation with commercial monocultures: annual crops, tree plantations of eucalyptus or softwoods, oilpalm, rubber and so on.

One means of funding biodiversity conservation and other environmental management has been the, sometimes controversial, debt-for-nature swap. Typically, a body like an international conservation NGO pays off some of the foreign debt for a country which otherwise might well default; in return the country undertakes to spend some of its national currency on conservation or other environmental works. In theory both the debtor nation and the environment benefit: the former saves face by not defaulting and has some control over the environmental activity, and the environment gets funding which would have been unlikely without the swap.

Also the NGO has promoted environmental care. Many conservation activities are expensive to establish, demanding land purchase, infrastructure, vehicles and so on; but cost much less for ongoing management, so foreign aid and measures like debt-for-nature swaps can provide crucial start-up money, and the developing country can then find recurrent funding. Even when satisfactory conservation areas are established, they are vulnerable to warfare, encroachment by

For some of these threats to biodiversity, simple legislation and conservation efforts may bring marked results. But when biodiversity damage is a side effect of economic development, controls may be more difficult. As prospecting and survey methods improve, and as demand from developed countries increases, mineral resource exploitation is likely to remain a challenge for environmental managers. There is also growing pressure for agricultural land and urban sprawl in many countries, which impact on biodiversity.

Developers can pursue the following resource exploitation options:

- Mining of non-renewable mineral resources: the source of income is exhausted, not sustained. Impacts include noise, pollution of streams and air pollution associated with cement production, lead, copper and tin smelting.
- Production of renewable mineral resources: this is potentially sustainable and includes sea-salts, sulphur from volcanic eruptions, nitrates.
- Sustainable yields of natural commodities from well-run plantations, farmland or natural vegetation: palm oil, natural rubber, rattan, bamboo, waxes, beverages, fibres and many other materials. This can be a problem if it takes land from smallholders and drives them to marginal areas where they can damage the environment. Commodity production may take land out of food production and make countries more reliant on importing grain. It can displace and destroy biodiversity, and can be a source of pollution as a consequence of processing.
- Unsustainable production of natural commodities: excessive hunting of species, over-collecting wild plants, clumsy ranching and plantations which degrade the land. Recently demand for herbal remedies and medicinal plants have led to over-collecting of wild biodiversity in many countries. Animals like tigers and rhino are suffering through hunting to supply traditional medicine. In Africa

squatters, fuelwood collectors, illicit miners, poachers and so on. Some reserves are too small or are in localities prone to environmental change and are not secure long-term.

Linking a number of such reserves with corridors, which enable species to disperse, can sometimes help. Another beneficial measure is to establish buffer zones around reserves to reduce external impacts on the actual conservation area. Although a reserve may be large enough to maintain biodiversity indefinitely and have enough resilience against expected climatic change, there are still disasters that can strike: bushfires, the arrival of pest species, strategic demands in time of strife and pollution from distant activities.

As India, China and other countries industrialise, acid pollution may well affect biodiversity in quite remote situations currently considered 'safe' and secure. Conservation efforts in New Zealand were recently endangered by one solitary 'eco-terrorist' who introduced predatory opossums to island bird reserves because for some reason he felt disgruntled.

Some countries and a number of development agencies and funding bodies are now insisting on biodiversity impact assessments for projects or policies deemed to pose a potential threat. Also, a number of environmental management systems have recently integrated biodiversity concern into their standards and prompted companies to publish biodiversity action plans outlining what they intend to do to improve things.

Impacts on Biodiversity

Developing countries commonly implement natural resource development projects, sometimes to generate export earnings. Sometimes there is direct mismanagement of biodiversity - forest clearance for timber, ranching or cropping; fisheries over-exploitation; extraction of wild products like rattan or ivory, wildlife trapping for the pet trade and so on. There are also indirect impacts, triggered by mineral exploitation, hydroelectric development, cement manufacture, industrial pollution, tourist disturbance and suchlike.

countries have an impact on poor nations as a result of new trends like buying exotic fruit, vegetables and flowers. They should be made more aware of this and of their obligations to pay something toward resolving any problems their consumer habits cause. Structural adjustment packages of financial assistance for countries in economic trouble have been agreed on the condition they impose financial austerity measures.

Bodies like the International Monetary Fund and World Bank have promoted structural adjustment since the 1980s, but the conditions they impose have had negative impacts on the environment of some countries. For example, farmers may have lost subsidies and turned to more exploitative land use off-farm, such as charcoal production, which damages the environment and wildlife. Another problem for biodiversity in developed and developing countries is the replacement of large swathes of natural vegetation with commercial monocultures: annual crops, tree plantations of eucalyptus or softwoods, oilpalm, rubber and so on.

One means of funding biodiversity conservation and other environmental management has been the, sometimes controversial, debt-for-nature swap. Typically, a body like an international conservation NGO pays off some of the foreign debt for a country which otherwise might well default; in return the country undertakes to spend some of its national currency on conservation or other environmental works. In theory both the debtor nation and the environment benefit: the former saves face by not defaulting and has some control over the environmental activity, and the environment gets funding which would have been unlikely without the swap.

Also the NGO has promoted environmental care. Many conservation activities are expensive to establish, demanding land purchase, infrastructure, vehicles and so on; but cost much less for ongoing management, so foreign aid and measures like debt-for-nature swaps can provide crucial start-up money, and the developing country can then find recurrent funding. Even when satisfactory conservation areas are established, they are vulnerable to warfare, encroachment by knowledge and skills, and ensure institution building provides stable foundations for supporting conservation.

Frequently there are excessive claims for the benefits of poorly managed ecotourism. There may also be a problem of seasonality, so that ecotourism only provides a satisfactory livelihood for part of the year. As with other forms of tourism, there is a certain amount of vulnerability to sudden changes in fashion or scares about travel caused by distant crises, which can suddenly hit profits and any environmental management which has come to depend on them.

Conservation can also be linked to other benefits; for example, a forest reserve may be managed to lock up atmospheric carbon to counter global warming or protect a catchment providing drinking water or flows to irrigation or hydroelectric plants. In steep areas vegetation cover is vital to prevent mass movement onto land below. There is also a crucial link with commercial research and development. A conservation project might even be integrated with several beneficial activities; an example in a developed country is the Great Barrier Reef World Heritage Area Strategic Development Project, which links conservation with over 60 organisations in reef-related food production, job creation, recreation, cultural heritage protection and so on.

Developed countries benefit from funding biodiversity conservation research and monitoring. Where there are commercial pressures leading to biodiversity losses, developed countries can support efforts to seek alternatives: for example, developing substitute products. Developed countries can fund advertising to counter fashions that endanger species. Treaties and controls can help reduce the trade in endangered species. Sales controls and taxation can reduce trade in endangered species and fund conservation activities, as can tourism taxes, airport taxes and levies on traded items - provided they are not judged to be a trade barrier. The processes of globalisation and free trade agreements need to be closely monitored to ensure they do not trigger biodiversity losses.

Nowadays, people are using more products derived from areas well away from where they live, so citizens in developed especially a wide variety of wild animals are being hunted to extinction to provide growing cities with 'bush meat' which is believed to have magical qualities as well as being a delicacy.

The unsustainable use of biodiversity and biodiversity 'theft' is nothing new; between the 1850s and 1880s English, Spanish, German and Dutch adventurers strove to steal seeds and plants of the best *Chinchona* species from Andean forests of Ecuador, Bolivia, Peru and Venezuela - the tree bark yielded between 1 per cent and 14 per cent quinine, which for nearly a century before chemical synthesis was the only effective treatment and prophylactic for malaria. *Chinchona* bark helped rid Europe of the disease and greatly aided colonisation of Asia, Africa and elsewhere.

Until biodiversity theft took place and the Dutch and English established plantations in the late nineteenth century the only source was from the wild and it was jealously controlled by native collectors, merchants and the governments of Bolivia, Columbia, Ecuador and Peru. The collectors failed to replant the stock they destroyed. So by the 1870s many of the best quinine-yielding species were on the verge of extinction. Those trying to steal for profit or to make cheap quinine available to the people of India could now claim they were conserving species going extinct in the wild. Natural rubber biodiversity theft, from Brazil in the early twentieth century by the British, was purely profit-driven or for strategic reasons. Once quinine and rubber had been removed the countries of origin suffered serious loss of revenue.

In the past profits have usually been taken from exploiting and damaging resources without any money put aside to rehabilitate the land afterwards - countries like Malaysia have a legacy of tin-mine craters and large areas of infertile sandy tailings (waste sediment). Some Pacific islands that were mined for phosphates lost their topsoil and now have a degraded flora and fauna; some islands were used for atomic weapons testing or as air bases; the latter in a number of cases lost biota as a consequence of rats, cats and snakes arriving.

Large mining operations and small bands of miners in various parts of the world spread diseases, hunt out game animals and pollute the environment with compounds like mercury which kill fish and other organisms even at low concentrations. Papua New Guinea has serious difficulties with the waste from large copper mines polluting rivers. Aluminium production generates problems associated with large-scale open-cast mining and demands large amounts of electricity for processing the ore. Many of the developing world's large dams service mineral extraction and processing, and cause serious environmental and socio-economic problems.

Tropical rivers have lost fish, aquatic mammals and other organisms because dams prevent migration, regulate flows and discourage spawning, or simply cause pollution and settle sediment, reducing downstream nutrient supplies. Floodlands and estuarine ecosystems suffered because dams reduce or stop seasonal flooding - which has tremendous impact on biodiversity. The solution is for dams to periodically release flows to replicate natural flooding.

However, managers are driven by short-term economic gain and hesitate because it would reduce electricity generation or irrigation storage - in a few countries, notably South Africa, artificial flooding regimes have been established. Tropical river pollution and disturbance by dams and other developments means that many potentially useful fish are lost before they can be assessed to see if they have aquaculture value, so the world is poorer. The problem is expressing the value of biodiversity in economic terms, in ways developers will heed. The cartoon book mentioned earlier in this chapter at least makes a start.

Ideally, a mining scheme should be required to store topsoil for rehabilitation and bank funds for restoration work, pollution control and compensation in the event of unwanted impacts. In time abandoned mining land can be of considerable conservation value even without restoration, especially if it is managed to encourage biodiversity. Unfortunately, countries are often keen for development projects to earn foreign exchange and provide employment and waive or weaken rehabilitation requirements.

Where land is scarce, abandoned land may be grabbed by speculators and squatters, rather than be left to harbour biodiversity. Restoration can also mean planting trees that are unsuitable for wildlife conservation, but are cost-effective and aesthetically acceptable, like eucalyptus. Environmental managers must explore practical ways of achieving 'responsible' mining and quarrying.

In some cases abandoned open-cast mines and quarry excavations can be used for refuse disposal - hopefully with suitable lining to prevent leached waste polluting groundwater or streams - aquaculture and recreation as well as (or combined with) wildlife conservation. The Eden Project in Cornwall (UK) has turned a disused quarry into a tourist attraction and important conservation facility which seeks to conserve tropical and subtropical biodiversity and offer other services such as research.

Wetlands - swamps, floodlands, estuarine marshlands, mangrove forests, lagoons and so on - are being lost at a worrying rate and are some of the most valuable environmental assets available to developing countries. Some of these are rich in biodiversity, and many also act as vital 'stepping stones' for migratory species. Damage to a few or even a single wetland may have serious impact on migratory species of birds and insects.

The main initiative seeking to protect wetlands is the 'Ramsar Convention', the Convention on Wetlands of International Importance Especially as Waterfowl Habitat. Many countries signed this in 1971, and it maintains a permanent bureau which works closely with the World Conservation Union (IUCN) and actively monitors and seeks to support wetland conservation.

Wetlands are widely settled and some of their people have developed some of the world's best-proven sustainable agriculture strategies which could have wider application.

programmes set up in the early 1980s in Japan and Europe, biotechnology research has not progressed as rapidly in the developing countries. There are a variety of reasons for this, the most important perhaps being a lack of funding for both basic and applied research and also a variety of restrictions on trade and capital investment, both of which have played important roles in the transfer of knowledge and capital equipment in the industrialised countries.

The notable exceptions among developing countries are the countries we know as the NICs, not only the four tigers, but also other countries in south-east Asia such as Indonesia and Malaysia which have made rapid strides in developing capabilities in the new biotechnologies by establishing multiple centres of excellence. According to the US Office of Technology Assessment (henceforth US OTA), Taiwan is perhaps best placed of all these countries in terms of absolute levels of resources to harness biotechnology.

As we shall see later in this chapter however, possessing resources itself may not be adequate for developing technological capabilities and transforming those capabilities into successful commercial ventures. Other countries in the region, such as Korea and Singapore, have also made great strides in developing modern biotechnology and have made significant investments in the US market which are being used to transfer knowledge and technology back to the home countries.

The second tier of developing countries which are making some progress in biotechnology research are those that have developed some capability in scientific research over the years but which are lagging behind the NICs. They lag behind mainly because of traditional bottlenecks in infrastructure, in economic and other regulations and, more often than not, in a lack of access to capital markets or other sources of finance. In Latin America this has been the main problem despite the fact that biotechnology policies were implemented in the early 1980s, shortly after a number of industrialised countries defined their own national programmes.

motor vehicles produced, there are considerable quantities of agrochemicals used during cultivation and waste generated during processing - pollutants which can decimate riverine species. Often, export crop processing is decentralised to ensure the produce does not deteriorate before it reaches the factory; consequently, if effluent controls are not effective, pollution is likely to be widespread. In Malaysia oil palm and rubber processing polluted a large number of streams in the 1960s and 1970s.

Aquaculture is often hailed as a route to improved and sustainable food production; in practice much of the huge expansion in developing countries is not sustainable and has marked off-site impacts. Several South American, Asian and Southeast Asian countries have serious problems as a result of developing lucrative tiger prawn aquaculture. Mangrove areas are destroyed to construct ponds, water resources are exploited and returned to the environment contaminated with waste and chemical pollutants. In order to feed the prawns inshore waters may be netted for juvenile fish and plankton, which also destroys wildlife.

Less environmentally damaging prawn and fish aquaculture strategies should be possible but require changed habits and investment. Presently developers can make big profits from unsound production methods and have little incentive to improve things. The developing country government is keen to support such production for foreign exchange earnings. The environmental manager faces a challenge - how to break such established malpractices and prompt a change to better approaches. Establishing a non-polluting intensive 'closed-cycle' prawn or fish aquaculture unit is still rare, and added to this is the need to find sustainable and unproblematic sources of stock and feed.

Where natural products are being overexploited the problem can sometimes be effectively addressed by imposing controls, such as extraction quotas, wardens, export restrictions, exploitation seasons, limited numbers of extraction licenses and compulsory methods; profits from the permits and charges can be ploughed into environmental management, but often get used for other things. It is also possible to establish protected areas where some biodiversity is retained but extraction is also allowed, or develop in-vitro production or substitution with an alternative product.

Natural resource production has generated considerable hindsight knowledge, so it should now be possible to plan future developments and retrofit those in place to cut unwanted impacts; the economics of production often discourage this. There has been considerable interest in using bioregional planning to support conservation. A bioregional approach promises to be a practical way of dealing with conservation and other environmental activities at a manageable scale.

Worldwide, since the 1970s, indigenous peoples have been establishing and exercising their rights over natural resources, notably in the USA, Canada, New Zealand, Brazil, Mexico and Australia. Some have made progress in improving livelihoods without untoward environmental damage. But the idea that indigenous peoples are always sympathetic to their environment, and so will not exploit it, is wishful thinking. There are unfortunately many cases of indigenous people causing environmental degradation, and sometimes central governments have found it difficult to intervene because rights have been granted.

If local people do not understand and support resource development controls it will be difficult to make these work: the locals know the land and its biota and if motivated make excellent wardens to police and manage conservation or other managed areas; if they are not involved their resistance is likely to be strong. A recent development - the extractive reserve - generally seeks to involve local people to combine conservation and sustainable exploitation with local livelihood provision.

The way forward for conserving much of the world's biodiversity *in situ* is through management partnerships involving individuals, communities, companies and governments (often in combination). Through these partnerships, collaboration can develop between those stakeholders wishing to exploit and those wishing to conserve.

Environment Trade Problems

Over the last 30 years or so a number of regional free trade agreements have been established, including the North American Free Trade Agreement, the European Union and the Association of South East Asian Nations. The principles of GATT were extended by the WTO in the 1990s, so that by the start of the millennium free trade rules determined by the WTO were effectively in force globally. GATT and the WTO had by 1994 laid down that there should be no discrimination or trade barriers. Ensuring that GATT does not cause environmental problems is the concern of the Committee on Trade and Environment of the WTO.

One WTO problem has been with the application of the precautionary principle. Because this is difficult to pin down legally, there has so far not been a full consensus on a definition, or whether or not to formally accept it. Consequently, there have been a number of situations where problems have arisen; for example, the WTO objected when Europe wished to restrict the import of growth hormone-fed beef, mainly from the USA. The WTO viewed this as a trade restriction because the precautionary principle was seen to be invalid by virtue of the threat being insufficiently proven.

A number of environmental management agreements have sought to use the precautionary principle; for example, the UN Biodiversity Convention and the Rio Declaration on Environment and Development hold that, even if full evidence is not available, where there is a serious threat, measures should be taken to minimise it. GATT Article XX (b) does allow measures such as quarantine for protecting human, animal or plant well being; but in practice the issue of justifiable risk is still problematic. Negotiations at Cancun in 2003 do not seem to have advanced things much.

There are international agreements that do reasonably effectively embrace the precautionary principle and try to fit in with WTO demands; the Kyoto Protocol of the Framework Convention on Climate Change, Article 2:3, for example, states that policies and measures seeking to reduce climate problems should be implemented in such ways as to minimise any adverse effects on international trade.

Opportunities for disagreement over trade and environment issues, especially relating to the precautionary principle, still abound. One example of trade barriers versus environmental protection would be where a state, unable to tell tins of tuna caught by dolphin-friendly methods from those containing fish landed in less environmentally welcome ways, imposes an import control on all tinned tuna. While this would help protect marine mammals it would probably infringe GATT rules. Some free trade measures conflict with health measures, like quarantine; trade in hazardous materials banned in some nations but not in others could hinder measures aimed at protecting the environment, like the Convention on International Trade in Endangered Species (CITES).

Since the early 1990s most of the former communist states have embraced some form of free-market economy. These transitional economies have shifted from centralised governmental control toward increasing activity by commerce. It is still not clear how effective environmental management will grow out of the confusion. Will the companies take the initiative? Will the state have to impose controls, or can there be a joint venture approach?

Energy Supply Threats

Compared with the 1960s and 1970s, worries about future energy supplies have been relegated to a side issue. There is ongoing research and development into clean sustainable energy sources; some would meet developing country needs, in particular appropriate and accessible technology, which is practical and affordable. Developing countries have been widely pressured or tempted to develop hydroelectric power to avoid petroleum imports and to keep down carbon emissions.

Given the numerous, often negative, socio-economic and environmental impacts, hydro-electricity generation is not as green as some try to make out. Worse, large amounts of methane are often emitted from large reservoirs and the irrigation development associated with some large dams is also likely to raise methane emissions - not be as green in terms of avoiding global warming as many think. Large dams are also prone to silting up, so are sometimes not the sustainable energy sources that they are made out to be.

A number of developing countries are effectively in a crisis situation with respect to fuel wood. The supply of fuel wood and charcoal can be major causes of environmental degradation, and the costs of purchasing the fuel can absorb a large amount of poor people's income. Finding alternative energy sources for millions of poor people scattered over wide areas is not easy. Providing sustainable supplies of fuel wood may be possible in some areas, through woodlots and hedgerows. Care must be taken to ensure the fuel-wood supply measures do not destroy biodiversity. Carbon emissions from burnt fuel wood would be at least part-balanced by woodlot regrowth.

In urban areas, sewage and refuse offer some opportunities for biogas generation, but smaller settlements may simply not have suitable waste or the resources to treat it. Solar power via photovoltaic panels, local wind turbines and mini- hydro plants, which generate less than 6 MW, could support village lighting, TV and small refrigeration units for healthcare or veterinary use. The challenge remains to find replacement fuels for cooking and heating.

Ageing

In developed and developing countries, age is becoming a matter of concern, especially in poor countries where there is unlikely to be much state or private provision for pensions. The proportion of elderly people surviving in developing countries is growing. In some countries the cost of care may compete with environmental management for funds. However, this group could also be a valuable resource which should be tapped to support environmental management; they might be used to help staff various initiatives, to educate children in green issues, or simply provide voting support if recruited to the cause.

Resistance and Resilience

A hazard is a potential threat; a threat is when there is reasonably clear danger. A risk is the probability that a hazard will happen or, more precisely, the likelihood that an event will coincide with elements which can be affected. There is some variation in definition and measurement of risk between disciplines: engineers, toxicologists, economists and ecologists all differ. One definition of risk is:

Hazard × vulnerability

However, an engineer might see it as:

probability × consequence

A toxicologist would probably define it in terms of:

threat > pathway > target characteristics

Some see risk in purely human well-being terms, others are also concerned for the environment. Risk management can be conducted at various times; often it is undertaken when a problem is manifest; there are also good arguments for it to be conducted in advance of any development, and for it to be repeated at regular intervals. Risk management can be subdivided into risk identification; risk assessment; risk perception assessment; risk communication; risk avoidance, mitigation or control; and emergency responses.

A risk assessment identifies, and measures, the significance and likelihood that a threat will happen. It is also likely to involve assessment of the vulnerability of humans, ecosystem or environmental features. Vulnerability can be defined as the degree of sensitivity to an impact; it need not imply awareness that the threat exists. Alternatively, it can be defined as the characteristics of a person or group in terms of their capacity to anticipate, cope with, resist and recover from the impact of a natural hazard.

Broadly, resistance is the ability to withstand a hazard; resilience is the ability to recover from it. Although there can be no precise definition, a disaster may be described as the realisation of a hazard with serious consequences, or the serious disruption of everyday patterns of life. Some authorities have tried to apply human or monetary benchmarks: for example, defining a disaster as when 'there are more than 100 dead'. Like threats, disasters can range from a very brief event like an explosion to something, like soil degradation, stretching over centuries or more.

Some risks are voluntary and some involuntary; sometimes threats are linked to opportunities. The situation is clarified by the following example: someone wishing to cross the Atlantic may seek to do so by rowing boat or ocean liner. The hazard is the same - a risk of drowning. However, the risk is much higher in the rowing boat.

The impact depends on the nature of the threat, on vulnerability and recovery, and is affected by resilience and availability of aid from unaffected areas. Many of the known hazards have been assigned scales of magnitude: for earthquakes the Richter and modified Mercalli scales, for winds the Beaufort scale, and so on. So a listing of the recurrence of disasters can be judged against benchmarks. For the public the most meaningful magnitude benchmarks are numbers killed or injured and cost.

Human activities can increase vulnerability to threats and may also alter their natural occurrence and generate 'man-made' hazards: for example, deforestation is likely to alter the runoff from a catchment, often increasing the risks of floods and silt deposition. Also, vegetational changes like forest clearance may trigger human disease outbreaks as people come into closer contact with wildlife reservoirs of infection and vector insects.

Threats may be natural, human-induced or often a complex mixture of both, made even more difficult to unravel by cumulative effects (more than one impact interacting) and by positive or negative feedbacks (processes which exaggerate or counteract a trend). Similar events at different points in time may set in motion different chains of causation; what happens once need not produce the same event the next time, so environmental managers must be prepared for surprises.

People differ in their perception of threats: they may range from disinterested or complacent, even in the face of significant risk of major hazard, to alert and reactive. So there are a number of possible reactions to a threat: it may be ignored; efforts might be made to avoid or mitigate it; or contingency plans and relief measures could be readied to help recovery.

Risk assessment and management deals with trade-offs: economic and political versus technological capability. Risk assessment is to some degree subjective and people may ignore warnings and judge things 'acceptable'; typically more risk will be taken if there is a benefit, especially a profit, to be made. Benchmarks used to assess risk include established standards like toxicity tests seldom made on humans, failures per kilometre travelled, and hindsight.

Environmental management must warn of threats and environmental limits which should not be exceeded, map the dangers and degree of vulnerability, identify development opportunities and suggest appropriate responses. Effective early warning demands the identification of potential threats and monitoring to spot if any relevant critical thresholds are being approached. It is now relatively easy to obtain maps and geographical information system (GIS) readouts of vulnerability to and threat of landslide, flood, avalanche, tsunami, drought and so on.

Warfare and civil unrest threaten environmental management and hinder sustainable development efforts in many countries - and problems may continue long after actual hostilities cease because people have been driven from their land, and weaponry like landmines may remain active. Chemical compounds like defoliant may pose an 'environmental' problem long after fighting has ceased; this was apparent after the Vietnam War, during which large areas of defoliated forest changed to rank grassland and never recovered because of soil and groundwater alterations caused by tree loss. Other areas of Vietnam were replanted with robusta coffee and are currently helping to drive down market prices and ruin many of the world's growers.

Unrest in many parts of the world, plus the impact of restrictions to control pollution and rising costs of pesticides, has meant the neglect of insect control. Consequently, locust damage and mosquito breeding have made a comeback in some developing countries since the 1970s - Mauritania announced a growing locust problem in 2004. The invasion of new areas by organisms which out-compete the established species can be natural because some physical change, biological adaptation or accident of nature aids dispersal, but is today more likely to be assisted by humans through tourism, business transport and hobby collectors of plants and animals.

Charles Elton warned of the threat in the 1950s and new examples increasingly occur. Sometimes the invasions pose a threat in the form of new or increased outbreaks of human disease, or they may threaten food production (Food and Agricultural Organisation); infrastructure through damage to channels and banks by burrowing crustaceans, molluscs or rodents; choking of pipes, canals and water bodies by weeds, molluscs or algae; or they may pose a nuisance. Many developing countries are very dependent on one or a few crops for export earnings and are thus vulnerable to introduced organisms. The move toward free trade means that effective plant and animal quarantine measures must be acceptable to General Agreement on Tariffs and Trade/World Trade Organisation (GATT/WTO) authorities and monitoring must be improved.

The bovine spongiform encephalopathy (BSE) and foot-and-mouth livestock health problems in the UK recently illustrate how easily outbreaks can occur and how costly they can be. The causative organism of foot-and-mouth had probably been introduced by migratory birds or the importation of uncertified meat. Warning of a threat may not be enough - the UK has long been aware of the dangers of foot-and-mouth - the problem is getting people and administrators to act in an appropriate manner to reduce the risks, and global monitoring to get early warnings. In the nineteenth century Europe nearly lost all its grapevines to introduced insect pests that transmitted disease. Cocoa producers are vulnerable to fungus diseases, and many other developing country crops are at risk.

Some threats, which cannot be prevented, might be controlled or induced in a way to cut damage compared with their unhindered effect; for example an avalanche can be deliberately set off before too much snow accumulates and at a time when people are evacuated. There are hopes that if the stresses causing earthquakes can be seen building up an area might be evacuated and then some means used to trigger it. Similar pre-emptive controls might be possible for dangerous gas emissions from lakes, and are already used against the threat of forest and bush fires and landslides.

Foreseeing threats in advance is often difficult; some 'creep up' in an insidious way - one moment there is no problem, and the next a full-blown disaster. The latter situation can give rise to a 'frog-in-a-kettle' scenario, where people experiencing gradually deteriorating conditions - for example, gradual soil erosion - may fail to act because there is no clear crisis point. A frog gradually warmed reputedly stays and boils, whilst one suddenly placed in hot water will try to leap free. Some threats are sudden and obvious, some slow and obvious, some slow and insidious, others sudden and insidious - the latter two are the least likely to provoke a response. An example of a relatively slow and insidious threat is human immunodeficiency virus/ acquired immune deficiency syndrome; an example of a sudden and obvious threat is the 1918-19 'Spanish' influenza pandemic. The latter is little remembered today but it killed around 22 million people - more than First World War military casualties.

Some very rare events can have catastrophic impacts, yet without historical experience people are unlikely to treat the threat seriously. Response depends largely on human perceptions of severity, extent and frequency of recurrence. Individuals, state agencies, media and non-governmental organisation (NGOs) may be able to educate people to respond to threats and can support monitoring and early-warning activities. For example, the FAO established the Global Information and Early Warning System (GIEWS) in 1971 to provide early warning if world food supplies looked likely to become problematic.

Japan, Hawaii and Pacific-coast USA have tsunami early-warning systems, and a number of countries have avalanche-alert systems. Weather forecasting has improved significantly in recent years, allowing improved hurricane warning, and river basin managers can make use of it to prepare for floods. Earthquake early warning is not yet developed to a useful degree and the same can be said for volcanic eruptions, although there are promising signs the latter might be developed.

Further complications can be identified. Different groups of people vary in their response to a threat: e.g. locals, regional administrators, state administrators, international agencies, world opinion, different age groups, gender groups, class groups, ethnic groups and others. Also, each may alter its response over time; consequently, people often react differently to successive similar threats. Some people are able to adapt, some panic, some are fatalistic or lazy and many may have too limited resources to make much response. Access to opportunities to overcome threats is crucial, and some groups have far more access than others, usually the rich. Where a single disaster may be withstood, repeated impacts of the same sort or an unfortunate combination of events may spell catastrophe even for a resilient group of people or organisms.

Poverty usually means people are condemned to situations in which they have little in the way of reserves to resist threats or to recover from them, and 'living near the edge' in such situations they are often very vulnerable. Administrators and politicians may often not respond to a threat because they are not well-enough informed; but sometimes it is because there are no funds, or action does not seem likely to win them promotion or votes. Forecasters and administrators run the risk of being seen to be 'crying wolf' - if they act proactively and then find the threat has not materialised, they lose face or worse and will be reluctant to act next time.

So if any action is taken the strategy increasingly adopted is to seek a win-win solution. This is a solution that has benefits if the threat becomes reality, and also offers something if it does not - hence it is usually a compromise, and not necessarily the most effective route. Another common response is to wait and see or at least delay to seek further proof and/or wait for technology to improve. Few modern administrators or politicians look further into the future than about five years; they generally show little concern for something 'not in my term of office'.

Unfortunately, many resource development and environmental management issues demand a much longer planning horizon. There is another unhelpful response that is quite common in developing countries and in some richer nations, which is for those active in environmental management or responsible for human welfare to have to report to headquarters and await permission to act. The delay may be long, and there may be reluctance to bring 'problems' to the attention of superiors for fear they are seen to be troublemakers, overzealous or unable to cope.

This top-down administration may have other impacts. It means centralised control which is likely to be out of touch; those with the education and skills to make wise responses may resist postings to remote sites and, understandably, in an uncomfortable and possibly dangerous environment, be reluctant to 'get their boots muddy'. There is a further possible problem: the developing country specialist may have been trained in a developed nation, and so is not sufficiently sensitive to issues. Where expatriate expertise is being used the same risk is present, coupled with the likelihood that they will move on before problems manifest without training locals. The Internet may help resolve these problems; as distance learning becomes available more developing country staff can be educated at a reduced cost.

However, an adequate fieldwork and practical component may be difficult to deliver with it. One reaction to a threat is to insure against it. Insurance spreads the risk; a large number of premium payers contribute manageable sums and, provided not too many make claims at one time they should get adequate compensation. Unfortunately, some threats materialise into problems, which affect many people at once, so that spreading the risk by insurance fails.

Insurance companies may counter by reducing payouts; re-insuring, so that the burden is spread, possibly worldwide; inserting exclusion clauses into cover; imposing excess charges; refusing high risk clients - but for some threats such measures cannot resolve things. Some insurance companies have great experience and have developed very effective risk assessment approaches, which can assist environmental management. In the UK, for instance, commercial flood risk maps have recently been made available to environmental planners.

Once a threat is realised, it may be possible to retrofit infrastructure or modify ongoing developments to reduce impacts and vulnerability. Adaptation to environmental and socio-economic threats is both a cultural (learning) and in the longer term a biological process. So education may reduce vulnerability and help prevent some problems. There may be situations where a disaster offers opportunities: depopulation through epidemic may enable land reforms or solve employment difficulties and prompt longer-term development, as seems to have been the case after the 13th century Black Death in Europe; without the upheaval, social, cultural and economic change would probably have been slower. Some disasters causing serious damage in the short term may offer opportunities over the longer term; for example, some volcanic ash-falls, which destroy crops and injure people, may weather to yield fertile soil. Conversely, disasters may trigger disruptive adaptations; for example, eco-refugees may spread and cause secondary impacts in a number of places.

Recently a lot of attention has focused on how humans affect the environment, especially through global warming; much less effort has been devoted to assessing how the environment affects humans. During the last few decades knowledge generated by palaeoecologists and environmental historians, plus the stimulus of some recent severe ENSO events like that of 1997-8, have helped stimulate interest in environment-human interrelations. Keys tries to convince his readers that there have been strong links between environmental events and human fortunes, singling out AD 535-6 as a global crisis point when a volcanic disaster triggered a chain of adverse weather, hunger and plagues.

However, while neo-environmental determinism makes fascinating reading and may offer valuable warnings, a good deal of it is not sufficiently thorough and objective. A historical correlation between, say, a volcanic eruption, evidence of failed harvests and social calamity, does not prove causation. There are very few indisputable correlations between environmental events and human fortunes, the AD 79 Vesuvius eruption being one; many others remain in dispute because some evidence suggests environmental and other socio-economic causation.

If natural disasters can be *reliably* proven to be periodic or quasi-periodic through research by environmental historians, synoptic climatologists and palaeoecologists, then administrators will take the threat seriously and useful forecasting is possible. That situation seems to be approaching with respect to ENSO events.

Thirty years ago, Meadows et al. fed information on what they saw as critical development issues (food supply, pollution, population growth and so on) into a computer systems model; the result was a prediction of future trends which showed crucial limits being exceeded. Their sequel re-examined the predictions and concluded a crisis would still happen within a generation or so unless environment and development was better controlled; recently others have explored 'future development crisis', one predicting serious trouble as soon as 2030.

Global Environmental Change

Considerable resources have been devoted to predicting the likely impacts of global environmental change, which is most widely assumed to be human-caused warming - the 'greenhouse effect'. A number of university and international institutes have modelled its probable impact on agriculture, human and livestock health, and other important issues. Global environmental change is multifaceted; there is ongoing natural change - random, quasi-periodic and periodic - and there is that partly or wholly caused by humans. Examples of natural change include the solar cycle, geomagnetism, orbital attitude (all possibly periodic); and volcanic and seismic events, asteroid and comet earth impacts, natural species extinction and evolution (no obviously repetitive pattern established).

Superimposed on natural changes are those caused by human activity: pollution, land clearance, hunting and so on. So far most human impact has been through hunting and pollution, but transportation is increasing the movement of organisms, and genetic engineering may become major causes of change. Some argue that human influences are now so great that a new current geological period should be recognised - the Anthropocene.

Greenhouse warming is presently seen as a 'key' environmental issue by many people and it enjoys huge media coverage. Developed and developing countries have been trying to assess likely impacts and predict future scenarios. Will richer nations be better able to withstand changes? Will the poorer be at a disadvantage, or will it be vice versa? Where will the greatest problems arise? Can developed and developing countries reach agreements to mitigate, avoid or adapt to threats?

The last 20 years have seen much-improved collection and sharing of data. While not by any means satisfactory there has been some progress in establishing talks and drafting agreements. Some states have begun to co-operate in lobbying for pollution controls and environmental change-related foreign aid. Assuming that the way global warming will proceed is uncertain, that the impacts are not easy to accurately predict, that change is already underway, and that pollution control agreements are relatively ineffectual - one could argue that environmental managers could achieve more by focusing on how to reduce people's vulnerability and improve adaptability.

Adaptability seems to have decreased in recent decades in rich and poor countries. Improving adaptability should be relatively easy compared with cutting global pollution, and it would serve against a wide range of unexpected natural and human disasters and environmental change. Environmental change may well have urban health impacts, but warnings are difficult to assess because many complex factors are often involved. Predictions so far are too often based on guesswork: for example, global warming is widely assumed to be likely to cause a spread and increased incidence of malaria. Yet in Europe, between the mid fourteenth century AD and the 1840s, malaria was common, although conditions were significantly cooler then than now. Factors other than climate seem to explain the demise of the infection in Europe.

One of the more likely impacts of global warming could be shifts in energy use and consumption of goods as rich and poor countries implement economic controls and introduce new technology to honour agreements over carbon emissions and adapt to climatic change. Environmental management must forecast and prepare for changes, and it must try to do so as objectively as possible.

Global environmental change is a transboundary issue - its causes are shared by all countries and its consequences will be faced by the whole world. In the last four decades there has been much more awareness of transboundary issues and the international impact of problems that would in the past have been mainly regarded as affecting only the poor of developing countries, e.g. multi-drug resistant tuberculosis.

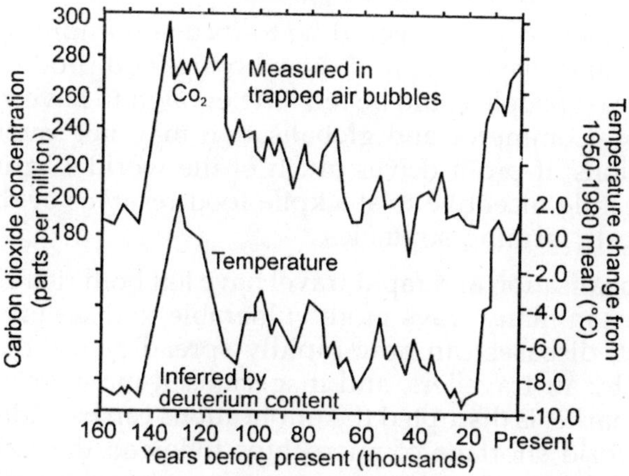

Fig. Global Warming

Greenhouse warming could have the opposite impact on some parts of the world; warmer tropics might mean more water vapour and greater snowfalls at the poles, cooling higher latitudes. Another risk is that warming will alter thermo-haline ocean circulation and in the Atlantic this might 'switch-off' the Gulf Stream and cause a sudden deterioration of Western European and possibly eastern USA climatic conditions. However, this has had little attention; most assume the future will be one of steady, manageable and predictable warming. Environmental managers concerned with development have to be aware of people's perception of threats, which may not be logical or predictable, and must weigh human vulnerability against threats they identify.

The currently dominant assumption or wishful thinking, that global warming will proceed in a gradual and steady manner, is unwise. Palaeoecology shows that there have been sudden thermo-haline circulation shifts in the Atlantic during the post-glacial period - the last 13,000 years or so - leading to sudden cold phases in Europe. A repeat would spell disaster to world food production, economic development and possibly peace. A wit once remarked that 'scientists and artists both tend to fall in love with their models'; many of the threats faced are very difficult or impossible to predict sufficiently accurately to give useful early warning, yet a lot of effort and funds are being spent on modelling.

A better response would be to focus on improving food reserves and dispersing and diversifying food production, so that at least people could be fed long enough to have a chance to adapt. Commerce and globalisation may not favour such precautions; if profit drives much of the world's agriculture, there is little incentive to stockpile food reserves or diversify small-scale farming strategies.

Globalisation and rapid travel have left both rich and poor countries in many ways more vulnerable to disasters than in the past: diseases can now rapidly spread across the world carried by air travellers, and if specialised manufacture for a global market is disrupted the implications can be widespread, e.g. a world shortage of microchips followed the 1999 Kobe Earthquake. In developing countries, rising populations, environmental degradation and erosion of social capital and traditional survival strategies have often increased vulnerability to threats.

Modern humans in rich and poor countries are less secure from environmental threats than they like to think and nature is not as benign and stable as most hope. Environmental management must assist governments to improve their citizens' adaptability and reduce vulnerability. Unfortunately, many anticipatory actions, and forward planning looking more than a few years ahead, get little support. Spending on improving adaptability and reducing vulnerability to a wide range of threats is wiser than expenditure on global warming controls.

Threats that recur, but only over hundreds or thousands of years, but which could be catastrophic, are ignored. This is unwise but understandable; people are reluctant to address threats that are seen to affect others at a distance in space or time. In spite of recent improvements in forecasting extreme environmental conditions, statistics indicate that growing numbers are killed by natural disasters in developing countries. This is in part due to population increase and settlement of more risk-prone areas as land becomes scarce, and also because poverty generally increases vulnerability.

The mid-twentieth to twenty-first century has witnessed the rapid growth of cities, especially in developing countries; large cities are complex and often vulnerable ecosystems, and once disrupted may recover with difficulty. Today's development projects, programmes and policies are often cumbersome and inflexible, and any change in human or environmental circumstances may cause considerable trouble.

For example, a large irrigation scheme may be constructed with little leeway for coping with altered runoff drainage in order to save funds, consequently even a small increase in storm intensity or alteration of crops or farming methods may cause problems. Technology may give new options but may also lock people into activities that pose a threat. Technology should be reviewed by environmental managers to try to ensure that it reduces human vulnerability. There may be situations where a threat demands a totalitarian response - faced with sudden and unexpected serious challenges, environmental management will have to be flexible.

Biotechnology Assessment

Before the introduction of a new biotechnological product or licensing of a new biotechnological production plant, its impact on the general welfare, health, economy, labour situation, culture and socioeconomic structures, etc. should be studied. Cary Fowler et al., 1988, Rural Advancement Fund International.

Biotechnology is a global issue. It cannot be assigned such attributes as positive, negative, or neutral. Like any other technology, it is inextricably linked to the society in which it is created and used, and will be as socially just or unjust as its milieu rational biotechnology policy must be geared to meet the real needs of the majority of the world's people and the creation of more equitable and self-reliant societies while in harmony with the environment.

The R&D and industrial sectors in biotechnology have aggressively sought product opportunities in the tradition of other high-tech ventures like microelectronics, computers, and robotics. But these industrial revolutions cannot compare to the commercialization of genetics in the public apprehension associated with their successes.

Geneticist Steve Gendel asks: "Why has biotechnology become such a focus for ethical, social, and economic debate while other technologies are all but ignored?" His answer focuses on the subject matter. "Clearly biological issues touch a sensitive aspect of our culture and leads to deeper and more passionate examination of issues than do issues raised by any other technology." I would argue that part of the difference lies in the fact that concurrent with the genetics revolution there have been challenges to the traditional ways of addressing the extemalities of industrialization. These challenges are confounding to government regulators and entrepreneurs who place their confidence in the established norms of social governance.

Industry and the regulatory sector share a commitment to incrementalism, although, by virtue of new legislation, the latter may be thrust into an alternative mode of handling the potentially adverse consequences of technological innovation. Established environmental laws and ethical traditions applied to biotechnology are being stretched to their limits.

The stresses have rekindled ideological debates on the responsibility of government toward new technologies. In this chapter I review some representative responses to the social management of technology. Then I explore the role technology assessment has played in evaluating a few high-profile products arising out of the new biotechnology industry. Finally, I outline a proposal for a more highly integrated approach to the review of new technologies that takes us beyond hazard control to social governance of innovation.

Environmental Traditionalists

Environmentalism, as distinguished from political and social ecology, is rooted in the constellation of laws that protect humans and segments of the ecosystem from the products and processes of industrialization. The vast majority of these laws that have been enacted at the federal level came in response to public concerns over the hazards of the chemical, nuclear, and fossil fuel industries.

Environmental traditionalists advocate a modification of the current regulatory system to address the problems of biotechnology. Some modifications, additions, and adaptations to the established regulatory regime of FIFRA, TSCA, and to a lesser degree the Food and Drug Acts, have already been made in response to biotechnology. The vast body of environmental law has not been amended by Congress. However, minor modifications of the existing statutory framework are well within the purview of the traditionalist response to the biotechnology revolution.

Reactionism

Among those who reject environmental traditionalism are individuals who advocate a libertarian model of technological innovation. According to this view, society should not assume the technology is hazardous before it is *proven* hazardous. Secondly, it is argued that the costs of pursuing "phantom hazards" are too great for society to bear. They cite ice minus as an example. It took five years and millions of dollars of regulatory review and litigation before an outdoor field test was permitted for an organism with a "mere" single gene deletion. The tradition of reactionism has attracted those who would eliminate the Delaney amendment for food additives, do risk-benefit balancing in assessing technological hazards, and place more emphasis on tort law and less on regulatory bureaucracy.

Social Ownership

Proponents of social ownership or social directorship of biotechnology argue their case from either a capitalist or socialist perspective. From the capitalist perspective, social investment should reap social benefit, while private investment should reap private benefit. Since the entire field of biotechnology arose directly from federal funding of molecular biology, under the logic of the economic system the public sector should be a key beneficiary in the outcome. In support of this view Barry Commoner stated: "We have to ask ourselves about the morality of allowing publicly produced knowledge to be taken over by the owners of capital."

This view is antithetical to the patenting of life-forms or the private appropriation of federally supported discoveries. From a socialist perspective, society will get the most out of biotechnology if its productive resources are directed by a state planning group or decentralized planning councils representing broad constituencies in society. Proponents of social ownership cite the direction that biotechnology takes under free market conditions. Profitability, and not social needs, dictates product development. Commoner, who advanced a similar argument for the direction of the energy industries, cited public control of technology at the sources of innovation and production as the solution.

"A fundamental question that any of us concerned with biotechnology have to deal with is the problem of governing the development of a new industry. I'm not talking about regulating its impact on the environment. I am talking about the social governance of the means of production." Without socially directed industrial development, Commoner and others argue, biotechnology will serve the interests of large established industrial corporations (petrochemical and agribusiness) and leave to pure chance the match between the productive capacity of the new technology and its contributions to the central problems of civilization (malnutrition, disease, environmental degradation, lack of inexpensive and clean sources of energy, prohibitively expensive health care).

Socialist solutions to the problems of postindustrial capitalism have lost much of their currency since the Reagan-Gorbachev era. With the world's major socialist economies (China and the USSR) exploring market alternatives, the rhetoric of centralized planning has far less appeal, even among democratic socialists. There is still much to be socialist about beyond the command economy and state ownership of the modes of production, particularly the public's role in determining the size and allocation of the federal budget for social needs.

But state economic socialism does not provide a sensible solution to harnessing biotechnology for the mases — at least not in the advanced capitalist nations.What alternatives are there beyond the three cited for the governance of biotechnology? I shall describe a system of social guidance that

I refer to as "market innovation-social selection." It is based on five premises.

1. The innovation sector and the social guidance sector shall be distinct. The main purpose of the former is to create new marketable ideas — to always be innovating — while the latter must evaluate these ideas within a highly articulated system of social directives.
2. The state shall expand its role in the assessment of new technologies. All new technologies must be evaluated on health and safety, ecological, equity, and ethical criteria.
3. Public participation in the assessment of new technologies shall involve all levels of political jurisdiction.
4. The state shall support maximum innovation in the private sector, but by a conscious process of selection, reinforce those innovations that meet important social needs and provide selective negative pressures against unneeded or unwanted innovations.
5. Only in cases where a robust system of private initiatives fails to meet public needs shall the state assume the role of innovator. However, in such cases (e.g., orphan drugs or recycling projects), innovation and social governance shall be the function of independent government bodies.

This system of social guidance for technology is modeled on Darwinian principles where two opposing processes (mutation and selection) provide the basis of growth, change, and balance. Innovation is essential for technological change.

But the state's role in selecting among competing technologies has been too limited and weak, and leaves too much to the control and self-interest of the innovation (production) sector. The current system is too product-centered. As a consequence it fails to account for technological directions. Social choices about the broad goals of technology are often the result of, or held hostage to, microeconomic decisions. The position I am advocating builds on a nascent form of technology assessment that began nearly two decades ago.

Technology Assessment

The Office of Technology Assessment (OTA) was created by an act of Congress in 1972. Under its congressional mandate, OTA's principal function is "to help legislative policymakers anticipate and plan for the consequences of technological changes and to examine the many ways, expected and unexpected, in which technology affects people's lives." Since its inception, OTA has published numerous studies and briefing papers on topics ranging from "Star Wars" to toxic wastes. In the arena of biotechnology, OTA has produced ten major studies and four background papers.

Despite the impressive volume of OTA studies in biotechnology, no generic framework of technology assessment links these diverse studies. Each assessment is carried out by a different staff, including a new project manager. Individuals who lead the studies are brought to the OTA for a few years and usually leave after their project has been completed. Topics for OTA analysis are proposed by congressional leaders such as the chairperson of a committee or subcommittee.

The OTA has developed its approach to technology assessment from a literature that has weathered 15 years of critical discussions. The end product for the OTA is a document that provides policy alternatives to Congress. For each study, the OTA convenes an advisory panel. The composition of the panel can affect how the assessment is carried out, which topics are emphasized, and which issues are muted.

A close examination of the studies reveals that technology assessment means many things to the OTA. To illustrate, in its study titled *Commercial Biotechnology*, the OTA placed considerable emphasis on what the federal government can or should do to protect America's competitive position in biotechnology. When the issue of environmental release of genetically modified organisms evoked public attention, a follow-up study gave primary attention to the ecological consequences of releasing novel organisms.

In applied genetics, the frameworks for technology assessment employed by governmental bodies often do not reflect the full scope of public discourse and critical debate. Primary emphasis is given to environmental and health impacts and/or to economic consequences both adverse and positive. This traditional orientation is grounded in a neoclassical paradigm that is comprised of several elements.

First, the private sector is primarily responsible for technological innovation. Second, the market system determines the social utility of the innovation. Third, the governmental sector protects society from the negative externalities of technology on human health or the environment. More recently a fourth component was added, and that is government's role in stimulating and protecting innovation in U.S. science and industry.

Starting in the mid- 1970s, a new critical school of technology assessment began to gain support, drawing its theoretical underpinnings from Lewis Mumford's humanistic sociology, decentrists like E. F. Schumacher, neo-anarchists like Murray Bookchin, and "soft energy" advocates like Amory Lovins. This movement rejected the neoclassical model of assessment and developed an alternative set of criteria against which to evaluate technology.

Factors like scale of technology, job displacement, ecological value, dependency on nonrenewable resources, quality of work, and degree to which the technology can be democratically managed became the new elements of an assessment programme. The concepts of efficiency and externalities that represented the marrow of the traditional paradigm were replaced with norms that encompassed broad social and cultural categories.

By recasting the parameters of technology assessment, critics of postmodern industrial capitalism expressed an allegiance to a new set of social values, such as deep ecology, the humanism of work, and the stewardship of nature. Technology became a vehicle through which to raise questions about the composition, control, and direction of America's industrial economy. Environmental interests merged with the interests of democratic decentrists and those who sought a stronger role for labour in setting economic policies.

Biotechnology is notable in that it is the first modern technological revolution that can be evaluated, in advance of its industrial deployment, by the technology assessment framework emerging from the new critical school. Despite this opportunity, little has been done to widen the lens of analysis for the first generation of biotechnology products.

When public issues around biotechnology are debated within the traditional assessment framework, much of the public discourse is left freestanding, without an institutional anchor. In some instances, issues raised by the public are viewed as unrealistic and dismissed as being naively romantic. To explore this point further I introduce seven factors for discussing new technologies: ecological impacts, human health effects, ethical soundness, economic productivity, distributive justice, social need, and market demand.

A product or new technology may thus be classified according to seven variables. This particular set of variables is useful in examining the current controversies over biotechnology. I shall use a ternary value system to distinguish different weightings of these variables: positive (+), mixed or uncertain (0), and negative (-). With a total of seven variables, each with one of three values (0, +, -), the possibilities are 3 or 2,187 cases. By assigning one of the three values to each of the variables, we can generate a specific assessment profile.

The distinction between social needs and desires provides a key demarcation between market and nonmarket approaches to technology assessment. In the former, desires are measured by consumer or market demand and needs are a function of desires. In other words, people need what they choose in the marketplace, and what they do not choose they do not need. There are, of course, always exceptions to the general rule. Poor people cannot fully express their needs in the marketplace. Therefore, they have needs that cannot be measured by consumer choice.

The critical school of technology assessment rejects the compression of needs and desires. It is acknowledged instead that something can be highly desired but not needed, or greatly needed and not desired. The orthodoxy of neoclassical economists, is blinded to the problem that wants are sometimes 'created by the production sector through advertising.

The fact that a firm develops and successfully markets a product is not *prima facie* evidence that the product meets an overall social need. Otherwise the following statement would be a tautology: any successfully marketed product produces an abundance of social utility over social disutility. The *informed* public response to a product or technological innovation provides a clearer indication of social needs than whether the innovation is efficient, profitable, or marketable.

In addition to social needs and market demand, the assessment criteria contain two biological factors (Ecological Impacts and Health Effects), two normative factors (Ethical Soundness and Distributive Justice), and an attribute that drives innovation and includes the profit element (Economic Productivity). The criterion of ethical soundness is an indicator of whether the product or technology is compatible or conflicts with existing ethical norms. Distributive justice refers to the impacts of the product or technology on the distribution of property/wealth. A negative consequence indicates a redistribution of resources in favor of the "best off" in society. The normative dilemmas of technologies are addressed in a variety of ways, sometimes within clear institutional parameters and sometimes outside of those parameters.

As examples, ethical problems associated with the application of human gene therapy are the responsibility of the NIH's Recombinant DNA Advisory Committee. Alternatively, ethical questions associated with creating transgenic animals have no natural venue within the federal regulatory system. Assessing technologies on the basis of their human health and ecological impacts rarely is an uncomplicated matter. Knowledge is often incomplete. Predictive models are seldom dependable.

Long-term effects are always problematic. Biotechnology, however, stands apart from other technological developments in one important respect: many of the questions about the impacts have been raised at the early stages of product development. This is characteristic of a social order that is responding to historical learning. Local communities that have questioned different facets of biotechnology responded out of their own experiences with technological risk.

But by raising the health and ecological safety issues early in the maturation of a technological revolution, the uncertainties are even greater. In some cases the risks can barely be articulated. The reaction to a new technology may result from a mistrust in technology per se, or there may be something about a particular technology or its product that strikes a raw nerve in a community. The novelty and unfamiliarity of biotechnology is certainly one factor that contributes to the public response. But that is only part of the equation. Some of the issues go much deeper to questions of control and accountability.

Traditional technology assessment has focused on the externalities of what are deemed socially useful products or processes. In theory at least, the constellation of environmental laws passed in the 1970s was designed to protect human health and natural resources. Among the most desirable and least controversial technologies are those considered socially useful (needed), sound on human health and ecological grounds, either ethically positive or at least ethically neutral, and favorable on considerations of equity. The zero values of EP and El indicate that the drug in question is not a big profit item and has no important ecological impacts (positive or negative).

Many public debates over technology arise because of the uncertainty inherent in explicating the health and environmental effects, regardless of the social utility. The early recombinant DNA debate serves as a useful example. In the mid1970s, most scientists did not question the social utility of recombinant DNA molecule technology. But they did question the safety of certain classes of experiments. An argument frequently heard is that society must always take risks if it wishes to support technological progress. Only a few scientists questioned the premise that genetic engineering meant progress, or the extravagant claims about progress.

During this early period, the scientific community was roughly divided into two camps, a division marked by the following query: Do we have sufficient information to insure confidence in the safety of transplanting genes from one species to another? The public debates that ensued did not accept those boundary conditions. They cited the ethical issues of human genetic engineering and the threat of biological weapons. Some individuals argued that we do not need rDNA technology or that there are moral reasons not to develop it. However, in the early stages of the rDNA controversy, the social utility of gene-splicing was not seriously questioned by more than a few faint voices in society.

In my criteria for technology assessment, I draw a distinction between a technology that is not needed and one that is not desired. To say that a technology or one of its products is not needed is to assert either that

1. We have an adequate substitute for it,
2. It is inconsequential (a marginal addition to the quality of life), or
3. It is detrimental to the physical, psychological, or social well-being of individuals.

A technology is undesired by some constituency when it is perceived to offer a greater balance of negative to positive utility. The public responds to undesired technologies exclusively through the marketplace. As an example, suppose a new technology is developed for sex selection of children. It may be argued that this technology is not needed by society (there are no sound reasons for selecting the sex of a child) and that it is also unethical as it may create imbalances in the world population or reinforce misogynic social mores. But this argument will not convince everyone and there will most assuredly be a demand for sex selection if it is available. Alternatively, there are technologies that some experts believe society needs but popular opinion is against, such as nuclear power.

3

Biotechnology and Communication

ROLE OF BIOLOGY

Information has provided a powerful metaphor for explaining the complexities of evolution and its complementary engines, DNA and environment; some of the theorems of information theory illuminate the constraints on the intergenerational stability of the genome. Dennett develops the information metaphor into a nontechnical explanation of the fundamental concepts of evolution of biological species, which he then extends to evolution of the ideas, practices, and patterns that constitute culture. However, Oyama claims that the application of information theory to evolution has sustained misleading dualisms between environment and heredity, material and form.

As a communication scholar, one is struck by the parallels to the ambiguous trace left by information theory on our understanding of symbolic communication between and among humans. In both cases information theory has literal and figurative applications to the phenomena of interest, and the conflation of a literal application of the mathematics of information theory with the underlying metaphoric understanding of certain complex phenomena in terms of signal transmission can lead the unwary into errors that obscure as much as they clarify. In Ritchie I show how the mathematical concept of statistical probability has become conflated with an epistemological concept of logical probability through a complex series of metaphorical mappings, and I use metaphor analysis to tease these concepts apart and explicate their relationship.

In Ritchie, I show how the interpretation of metaphor often requires a consideration of the broader sociocultural context as well as of the immediate textual context: In some cases expanding the context within which a metaphor is interpreted leads to a radical reformulation of meaning. In this chapter, I propose to apply a similar analytic technique to teasing apart the literal and metaphorical uses of information and information theory in writings about biology, and show how the conflation of the literal and metaphorical in writings about biology parallel the conflation of literal and metaphorical understanding of information theory in writings about human communication. I begin with a brief review of Shannon's basic ideas, including his overt use of certain compelling metaphors as a way to explain abstract mathematical ideas. I then summarize Dennett's use of information theory to explain evolution theory and Oyama's critique of the use of information theory in genetics. Finally, I return to an assessment of the interplay between the literal and metaphorical applications of information theory.

METAPHOR

Lakoff and Johnson argue that conceptual thought is fundamentally metaphorical. They have shown how our understanding of even the most basic conditions of our existence can be traced to metaphors grounded in our experience of the physical conditions of life. For example, they identify a family of orientational metaphors ("The stock market is up ten points, " "I'm feeling low today," "I felt a rising sense of anger and shame") and show how these metaphors link abstract ideas to embodied experiences such as sleep and waking, illness and health, emptiness and fullness. Lakoff and Nunez extend this reasoning in an ambitious attempt to show how even the most abstract forms of mathematical reasoning develop as a chain of metaphoric extensions from a small handful of innate perceptual abilities.

Vervaeke and Kennedy criticize the concept of implicit metaphors, contending that the identification of root metaphors in the interpretive process is indeterminate, and that root metaphors are implicit—not in the sense of inevitability or uniqueness, but rather "in the sense of 'waiting' or 'available' or 'apt once they are mentioned'"

According to Vervaeke and Kennedy, metaphors do not necessarily constitute thought, but are chosen from an indeterminate range of possibilities to express literal ideas. A set of metaphoric statements may be consistent with any of a number of root metaphors, each with its own distinct set of entailments, and the selection of any particular root metaphor for analysis will almost always be underdetermined. Thus, multiple interpretations of the same body of expressions are often plausible, and metaphors may be chosen deliberately to express particularly complex or subtle ideas.

Vervaeke and Kennedy also point out that novel metaphors can be produced by extending an identified root metaphor, but the novel metaphors thus produced "may or may not be apt or may run counter to the examples in the original corpus". Not all the entailments of a root metaphor can necessarily be mapped onto the target concept. As a consequence, even an interesting and informative metaphor may fail to generalize, and metaphor analysis is unlikely to lead to certainty about underlying meanings.

I have elsewhere extended Vervaeke and Kennedy's argument by suggesting that the analyst may usefully consider more than one root metaphor while trying to make sense of discourse. Metaphor analysis can help unpack implications of the language used in discussion, and thus broaden our understanding of underlying concepts. The analyst must carefully consider entailments of a proposed root metaphor, rejecting those that seem contradictory with the intent of those who use the metaphor as well as those that simply do not fit and accepting only those that convincingly contribute to understanding the target concept.

Kittay mentions a set of expressions that may be taken either literally or metaphorically. For example, Kittay cites a prisoner explaining why he did not escape by saying, "My hands were tied. I could do nothing. " Or we might say of an overweight friend who has chanced into a sinecure, "Fred is fat and happy," or laughingly say of a grossly exuberant friend, "Tom is an animal." Whether the intention is literal or metaphorical can be determined only by the context. In Ritchie, I extend this idea further, showing how the interpretation of metaphorical/literal expressions may depend not merely on the immediate context, but on the extended context that includes social, cultural, and political assumptions and beliefs.

For example, Lakoff and Turner analyze common euphemisms for death, such as "passed away," "was taken from us," in terms of root metaphors, "Life Is Being Here" and "Death Is Departure." But many religious traditions identify the person with an immaterial soul, distinct from the body, and regard death as literally both a departure and liberation. Moreover, for a person familiar with such a religious tradition (such as a person raised in but no longer belonging to a conservative religious group), the usage may be metaphorical, but based on the root metaphor, "Death Is The Departure Of An Immaterial Soul." The use of one of these expressions cannot be understood without considering the cultural, religious, and sometimes even political contexts.

In this chapter, I extend this line of inquiry further by showing how the proposed approach to metaphor analysis can help resolve ambiguities resulting from a conflation of literal with metaphoric entailments of complex ideas such as information. Information, information theory, and several related concepts are often used in a way that, like "fat and happy," "animal," or "passed away," can be understood as literal, metaphorical, or both at once depending on the context.

A further consequence of Vervaeke and Kennedy's argument is that both the rhetor and the rhetorical analyst must consider metaphors and the relationship between metaphoric and literal meanings in a deliberate way. Primary tasks of metaphor analysis when such complex metaphorical usages are involved include opposing the conflation of the metaphorical with the literal by first teasing apart the metaphorical and literal meanings and then separating apt entailments of the root metaphor(s) from inapt entailments.

Information Theory

Information theory has been understood with reference to several distinct aspects of Shannon's work as well as a broad array of more or less related work. As developed by Shannon, information theory began as an attempt to describe the constraints on transmitting signals by way of a medium (notably an electronic circuit) of known capacity. In this sense, *information theory* refers to Shannon's theorems stating the relationship among channel capacity, transmission rate, and accuracy.

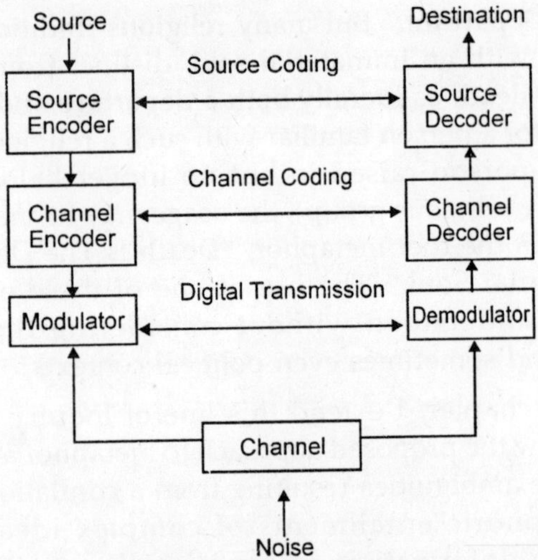

Fig. Information theory

More specifically, people have sometimes identified information theory with Shannon's expression for the variance or "scatter" within a set of elements such as a code or alphabet, in terms of binary digits or "bits, " H = -kóp(i)$\log_2$ p(i). This formula for H, which is virtually identical with the formula for entropy, plays a central role in Shannon's development of the theorems (discussed later) and thus lies at the heart of information theory. It has also come to serve, especially in the field of Communication Studies, as a powerful icon that seems to guarantee prestige and scientific status to any text in which it appears.

Sometimes Shannon's famous model of a transmission circuit (source message-receiver) is understood as information theory. Recently, the much more general theory of digital computation (information processing), made possible in part by Shannon's original formulation, has also been taken to constitute information theory. In general, any discussion of factors affecting the organization, description, transmission, or replication of patterns, such as the various formulations of chaos theory and self-organizing complexity, may also present claims for inclusion within the general meaning of *information theory*. Here I focus primarily on the model of a signal transmission system and the theorems that describe the constraints on transmission through such a system.

It is worth recalling the problem Shannon originally addresses, which was to develop a theoretical basis for increasing the efficiency of electronic circuits. Although Shannon was an employee of Bell Labs, it is clear throughout his article that he intended the theory to have application well beyond telephone circuits, including radio and TV broadcasts and indeed any medium of signal transmission whatsoever. Indeed by titling his article "The Mathematical Theory of Communication" and by drawing illustrative examples from such diverse activities as cryptography, the structure of language, and crossword puzzles, Shannon explicitly encourages a broad application of his work.

Electronic signals are subject to random perturbations, metaphorically referred to as *noise* because of the way human beings subjectively experience the sounds produced in a telephone or radio receiver by such perturbations. The second law of thermodynamics, famous as the "law of entropy, " implies that such random perturbations are both unavoidable and unpredictable because electronic signals are a highly organized form of energy, and energy tends to degrade from a more ordered to a less ordered form. However, the effects of these random perturbations can be detected and counteracted through various forms of redundancy—for example, by repeating the entire message or inserting check digits at the end of each string within a message. Part of the question for Shannon, then, was to formulate measures for the maximum capacity of a circuit, the degree to which random perturbations reduce that capacity, and the limits to the improvement of accuracy and transmission rate that can be achieved by introducing redundancy into the code.

Any message can be transmitted through any medium by encoding it into a series of detectable variations. For example, the telegraph relies on patterned interruptions in an electrical current; the telephone and radio encode the analog qualities of the human voice into analog electromagnetic waves. (Digital transmission encodes the voice into sequences of digits that characterize the amplitude of the sonic wave at regular intervals several times per second.) Semaphores create patterns of flags; Paul Revere's famous code used a pattern of lighted lanterns. The human voice encodes words into patterns of sound waves; alphabetic writing further encodes each discrete sound into patterns of letters.

In the simplest form of medium, such as a telegraph wire, Paul Revere's lanterns, or the bonfires by which news of the fall of Troy was spread, the code is digital. Familiar contemporary examples of digital codes include Morse code and the three taps on the ceiling or two bangs on a radiator pipe in the old country-western song. Reflection on the nature of messages leads to the realization that they always take a form very much like a formal code.

In particular, there must be variation (if everyone went around humming middle C, there would be no such thing as language), but there must also be organization (if everyone went around making a perfectly random series of sounds, there would be no such thing as language). The set of distinguishable elements, along with the rules by which they are matched with elements from any other set and assembled into messages and the rules specifying the internal structure of allowable messages, constitutes the code.

Digital codes such as Morse code suggested a ready metric for the capacity of any code: How many binary digits (e.g., zeros and ones, on and off states, bangs on the radiator pipe) are required to specify uniquely any particular element in the code? This problem is metaphorically similar to the problem in thermodynamics of uniquely specifying a particular energy state of a system (e.g., the dispersion of molecules in a container of gas).

That is to say, the problem of measuring the capacity of a code is metaphorically related to the problem of measuring the entropy of a self-contained physical system. Gibbs' formula for the entropy of a physical system provided a convenient and intuitively apt expression for the amount of variation among elements of a set, so Shannon adapted it to his own needs. The first part of his paper is devoted to showing that a modified version of Gibbs' formula (H, given earlier) is suitable to the task. Indeed so suitable is it that it is difficult to imagine a better measure. However, Shannon explicitly refrained from claiming that H is a unique or even the best possible measure. He simply shows that it is adequate to the task.

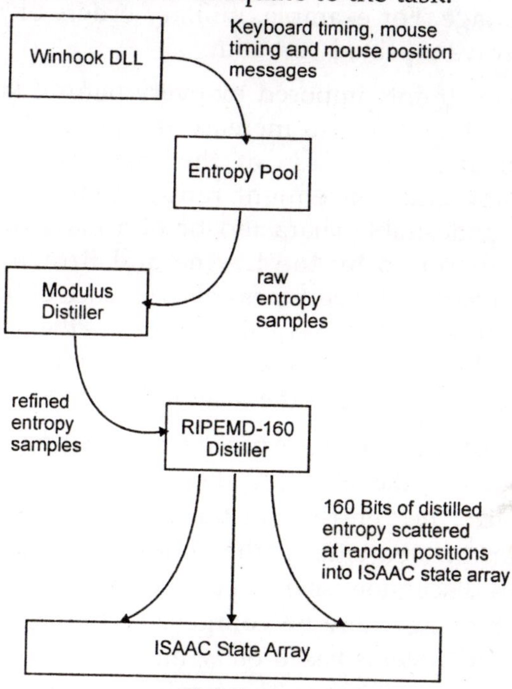

Fig. Entropy

When Shannon uses the word *entropy*, as in "entropy of the signal" or "entropy of the source, " he always refers to the degree to which elements of the source or signal are scattered. The value of H, as a measure of information capacity of a code or transmission channel or as a measure of the amount of channel capacity required to transmit a certain message, reaches its maximum when the distribution of elements (e.g., dashes and dots, letters of the alphabet) is perfectly random (i.e., when at any point in a transmission each element is equally probable).

However, a perfectly random distribution of elements has the undesirable characteristic that, if a random event (i.e., literal physical entropy) changes one element of the code, the error is undetectable and consequently uncorrectable. Error detection and correction can only be accomplished if redundancy has been introduced into the code. Redundancy can take various forms; the simplest form is merely to repeat every message. For example, military orders or reports are almost always repeated verbatim.

The constraints imposed by every natural language on spelling and syntax also increase the possibility for error detection and correction at the expense of increased redundancy and consequent reduced channel capacity. (Another undesirable characteristic of a maximum capacity code is illustrated by the Levine and Brown 1970 song mentioned earlier: If every possible sequence of bangs on the radiator pipe is used with equal frequency in the code, it would be impossible to distinguish an intentional message from simple malfunctioning of the heating system.)

Shannon's theorems are based on comparing the "entropy of the source, " the statistical dispersion of elements in the source, with the "entropy of the channel, " the statistical dispersion of elements in the code. Throughout most of Shannon's discussion, *source* refers to the set of elements of which a message may be composed. At the simplest, for transmission systems based on written English, the source is the letters of the alphabet. The maximum value of H—that is to say, the maximum capacity of the alphabet as a code—would be achieved if every letter appeared with equal frequency at every point in any message; in other words, if the assembly of a message were unconstrained by rules of spelling and syntax. But in fact, some letters (such as e, a, and s) appear much more frequently than others (such as x, z, and q).

Moreover, if a message includes the string "th," the next letter is very likely to be a vowel and will virtually never be "z" or "n." Expand the preceding string to fewer than a dozen letters and the next letter is often all but certain. Readers make automatic and usually subconscious use of this redundancy: When we come across a phrase such as "The chances are quite thn," we insert an "i," reading the phrase as "The chances are quite thin." Because of the redundancy built into the language, nothing else makes sense. Often the error correction happens subliminally—a fact that can pose a great challenge to proofreaders who may see the word as *thin,* with no awareness that the visual circuits in their brains have filled in a missing "i."

It is possible, by studying the way letters are combined into words and words into phrases within a given language community, to detect patterns such as these, which are repeated with sufficient frequency that channel capacity can be saved by encoding each of these patterns as a unit. Thus, when developing a code for transmitting news stories, entire stock phrases (stereotypes) may be treated as a unit and assigned a single sequence of digits in the transmission code.

A code might even be designed to permit maximum efficiency in transmitting messages from a single speaker or author (e.g., the president of the United States). So the source can refer to the collection of messages typically sent by a language community or even by a particular writer or speaker. Analysts of literary style sometimes use this technique, for example, to identify the author of a disputed text by comparing the conditional frequencies of elements in the text with conditional frequencies of the same elements in authors' known works.

The important thing to realise is that *source,* as used in Shannon's theorems, is distinct from the use of the word *source* as in "the source of a quote" or "a newsworthy source." When information theory is applied to genetics, *source* in this sense may refer to the collection of four nucleic acids or the collection of genes (patterned sequences of the four nucleic acids) that appear in a particular genome. The chromosomes of one particular parent then constitute a message composed of units drawn from that source.

Note that the relationship of H as a measure of transmission capacity to entropy, like the relationship of signal perturbations to noise, is primarily metaphorical even as it is based on a literal linkage. Entropy literally affects signal transmission inasmuch as signals are always transmitted in the form of organized energy states in a self-contained physical system, and these organized energy states are always subject to degradation through random physical events. (For example, the pattern of ink on this page is subject to alteration by the passage of ions through the atmosphere, by the abrasion of readers' fingers, and by spills of coffee, Coke, or wine on the page.)

However, codes and messages are abstract entities distinct from the physical systems described by the laws of thermodynamics: Describing the former in terms of the latter is unavoidably metaphorical. Messages must be stored and transmitted by way of patterns in a physical medium, but the patterns are distinct from the medium in which they are expressed. If the medium is altered in some way (e.g., if lightning strikes a telegraph wire), the observed pattern may be altered as a result, and the message may be partially altered or lost. H, as a measure of the *entropy* of a source or message, refers to the degree of scatter or apparent randomness in the distribution of elements within the source or message.

As a measure of the "*entropy*" of a channel, H refers to the degree of scatter or apparent randomness in the distribution of elements in the code by which messages are transmitted. H has only a metaphorical relationship to physical entropy—the tendency, articulated in the Second Law of Thermodynamics, of all energy systems to become more disorganized. Literal physical entropy affects the transmission system by creating the randomizing noise that introduces errors into transmitted messages. Thus, as a measure of the noise in a channel, H refers to the probability that any one element of a transmitted message will be altered because of random events affecting the transmission.

According to Zencey, entropy has become a widespread and popular metaphor for the relationship between order and disorder often applied in areas such as history, morality, and even religion, in which thermodynamics has little if any apparent relevance. It is true that all of these events involve physical entities in some way, and all physical entities are subject to the laws of thermodynamics.

The random events described by the laws of thermodynamics could conceivably lead over time to the breakdown of religion and morality—but entropy in this literal, physical sense rarely plays any serious role in theories of religion, morality, or statecraft. It is rare that enthusiasts for the concept of entropy take the trouble to show how physical entropy might be related, causally or otherwise, to the examples of disorder that concern them: The usage is virtually always metaphorical. Not surprisingly, a similar fate has befallen information theory and the concepts associated with it, including information, noise, and redundancy.

Unfortunately, metaphorical abuse of information theory was licensed almost from the outset by Weaver's interpretive extension of the theory and by the advice given by prestigious scholars such as Wilbur Schramm to "read Weaver first." It did not help that many students and scholars of human communication have lacked the mathematical sophistication to understand Shannon's original formulations and so never got around to reading Shannon at all. Consequently, information, like entropy, mutated into a quasimystical concept empowered by the aura of apparently impenetrable mathematics.

Indeed the two concepts are frequently linked—for example when information is equated with negative entropy and credited with the power to neutralize the unpleasant consequences of the second law of thermodynamics, such as old age and even the eventual descent of the universe into total disorder. (The right information could, of course, be used to counteract entropy and slow or even reverse the aging process within the bounds of a single organism, but it would be at the cost of a much greater increase in the entropy of the surrounding environment.)

It is useful to remember how much Shannon draws on other metaphors in developing, explaining, and justifying his ideas. He explains the problem of specifying a particular message out of a group of potential messages and related ideas such as redundancy in terms of several more or less familiar processes including cryptography, linguistics, and crossword puzzles. The word *information* is metaphorical—a metaphor reinforced by Shannon's extensive use of everyday human communication activities such as crossword puzzles (and not so everyday activities such as cryptography) to illustrate his concepts.

"Inform" comes from the Latin verb *informare,* to inform or give form to; *informare* is derived from *in-* plus *forma,* form. Most directly, then, *information* is that which informs or gives form to, or the process of informing or giving form to. The information capacity of a circuit or code (what H measures) can be understood metaphorically as the capacity of messages formulated in the code and transmitted via the circuit to create form (through the decoding process). What is "informed" is unspecified: Shannon explicitly disavows any interest in the meaning of the messages transmitted (the "forms" resulting from the messages), and later expresses considerable concern over the casual metaphorical extension of his ideas to the realm of human communication and to meanings.

In the study of human communication, information theory has proved to be a singularly dry well, at best evocative, at worst downright misleading. The real news of Shannon's work, which has been largely ignored by scholars of human communication, is in his theorems that specify the relationships among the complexity of a code, the capacity of a channel, the degree to which the channel is subject to random perturbations (noise), and the amount of redundancy built into the code. Communication is always based on transmission of some sort of signal, whether it be spoken or written language, a nonverbal gesture such as a touch or nod, or, perhaps, brainwaves or "psi" energy. Because transmission of any sort of signal is subject to random degradation, any communication channel whatsoever is a "channel with noise."

Based on these considerations, Shannon demonstrates that

1. Perfect, error-free transmission is unachievable;
2. At least under some circumstances transmission can be brought arbitrarily close to perfection; but
3. Incremental decreases in transmission errors are costly in terms of increased channel capacity, decreased rate of transmission, or both. To see why this is so, consider again the military rule that every order must be repeated: There is a finite, if small, probability that a randomizing process will produce exactly the same error in both versions.

Another form of redundancy is the use of a "check digit," representing the sum of all digits in a sequence of specified length. If one random event alters one of the digits in a string, another random event can alter the check digit in such a way that it appears to add up. In general, no matter how cleverly the code is constructed, no matter how many error-detection features are built into the code, it is always possible that some sequence of random events can both produce an error and alter the error-detection part of the code so as to produce the appearance of an error-free transmission.

Thus, in a universe in which random events occur, error-free transmission is impossible. However, it is possible, by introducing additional error-detection features into a code, to create a situation in which a huge number of improbable coincidences must occur for an error to slip by undetected.

Thus, it is possible, through added redundancy, to reduce the rate of actual errors until it is arbitrarily near zero (but never equal to zero). Finally, every form of redundancy requires its own part of the signal, and thus reduces the amount of the signal left for transmitting the message. Each increase in accuracy of transmission carries the cost of a decreased rate of transmission.

With respect to human communication, it follows that if perfect transmission of signals is unachievable, perfect inter subjectivity is unachievable. Any theory that relies on more than an approximate inter subjectivity is fatally flawed. To most observers of human communication, it probably comes as no news that perfect intersubjectivity is unachievable. But had theorists taken seriously the true content of Shannon's theory, that is to say the theorems, a long and fruitless theoretical debate over intersubjectivity might have been avoided.

Incidentally, similar reasoning applies to the research process. A researcher observing a phenomenon of any sort is detecting signals: Both the signals and the process of detecting the signals are subject to random perturbations. Although it is in concept possible, through instrumentation and methodological improvements, to come arbitrarily close to perfect reception, absolute perfection can never be achieved. Thus, true certainty is unachievable.

However, in both instances, it must be noted that the major sources of erroneous understanding are in the processes of interpretation, not in reception. Pending major breakthroughs in brain science, information theory can only be applied metaphorically to the processes of interpretation. To return to the example of military orders, the greatest source of error in orders is faulty reasoning on the part of the officer giving the orders or preconceptions on the part of the person receiving the order.

In the first place, if a bad order is given once, it will almost certainly be given the second time. In the second place, if a preconceived expectation as to what the superior officer is likely to command leads the subordinate to mishear or misinterpret an order, no amount of redundancy can ensure that the mistake is corrected. The movie *Crimson Tide* illustrates the problem of interpretation quite well, when the overtrained captain misinterprets a partially received "stand-down" order as mere noise and nearly starts a nuclear war despite the interpretive redundancy built into the dual-command structure.

Although information theory describes the absolute limits to intersubjectivity, on the one hand, and to scientific discovery, on the other, the greater uncertainties associated with interpretive processes keep us so far inside those limits that information theory is unlikely to lead to any truly useful insights.

Application to human communication is based on an implicit metaphor, "discourse as signal transmission," and application to epistemology is based on an implicit metaphor, "nature as transmitter." Both metaphors require careful inspection and justification before application of Shannon's theorems can be accepted. Signal transmission is certainly part of discourse, which requires transmission of messages in some physical medium, but it is commonly a very small and ordinarily the least significant part.

Theory and Evolution

Although other writers have applied information theory to biology, in both literal and metaphorical senses, Dennett's account of evolution theory provides a particularly useful example. Dennett sets out to generalize the ideas embodied in evolution theory to other branches of knowledge, in particular to human culture. Dennett develops a concept of *design space* based on Argentine poet Jorge Luis Borges' account of the "Library of Babel." The Library of Babel contains all the information in the universe, which is for computational (and bibliographical) convenience organized into books of 500 pages each, with 40 lines of 50 spaces on each page. Each space is printed with 1 of 100 possible characters (including spaces, punctuation marks, etc.).

If every possible book is included in the Library of Babel, there will be 1001,000,000 books in the library, including, for example, 100,000,000 versions of *Moby Dick* that differ from the canonical version by a single typographical error. The Library of Mendel, consisting of every possible combination of the four nucleic acids, forms a subset of the Library of Babel.

There are in the Library of Babel an uncountably large number of brilliant novels and poems, most of which will never be written, an uncountably large number of brilliant symphonies, most of which will never be composed; and still a larger number of banal novels, poems, and symphonies, most of which will also (fortunately) never be written. In the Library of Mendel, there are an uncountably large number of genomes specifying potentially viable organisms, most of which will never appear on earth. Each potential masterpiece, each potential work of pure kitsch, and each potentially viable genome is, like *Moby Dick,* surrounded by a sea of near misses. By far the greatest number of volumes in the library consists of pure nonsense.

This metaphor serves to render the problem of how a particular new organism could evolve from a particular existing organism to one of how random changes in the existing gènome could produce a genotype that is also viable, and how a series of such random changes could lead to a particular new genotype. If any link in the chain does not specify a viable organism, then the sequence of mutations is impossible. For any new organism to evolve, there must be a path that leads from an existing organism through a series of survivable organisms.

The environmental influences on an organism's development can also be specified, and the environment(s) in which a particular genotype can produce a viable organism can be incorporated into the description of that organism in the Library of Mendel. The logic of the metaphor does not change.

For example, several species of organisms can survive in extreme environments such as an atmosphere dominated by methane. For such an organism + environment to evolve from any existing organism + environment, it was necessary that a series of changes in the genomes specifying both the subject organism and the other organisms involved in maintaining the pertinent environment occurred, and, at each stage of the sequence, each of the organisms specified by these genomes had to be capable of surviving and reproducing in the environment they collectively produced and maintained. Otherwise, "you can't get there from here."

Although it is not central to the story I wish to tell here, it is worth noting the consequence for Dennett's enterprise of incorporating environment into the digital specification of an organism. The survivability of the organisms specified by human DNA depends, as Dennett shows, on a cultural as well as a physical environment. (Human infants cannot survive to sexual maturity without a nurturing and protective cultural environment.) Moreover, the organization of any individual's brain cells and their states at any given moment can be specified by a subset of the information in the Library of Babel—a subset that includes the individual's DNA as well as the accumulated physical and cultural experiences of a lifetime.

Through this line of reasoning, Dennett proposed to show that a potentially meaningful cultural product can actually occur only if there is a viable bridge from what currently exists (in DNA plus physical environment plus cultural environment) to the potential product. Many of the potential masterpieces will never be produced because the person who might have produced them will take or has already taken a different path. Poet Gary Snyder's sensibilities, encoded in the cells of his physical body as well as in his written works and other artifacts, are unalterably affected by his early study of Native American lore, by his studies in Zen, and by his life in a remote mountain community.

The poetry he might have written had he chosen to study Catholic or Sufi mysticism rather than Zen can no longer be written. What we have is very valuable; what we might have had can never be known: You can't get there from here. Explicit in Dennett's entire argument is a fairly sophisticated rendering of both culture and biology in terms of information theory.

DNA is a code; genes are digits in the code—the equivalent of the English alphabet. The cell—starting with the ovum—is a genetic *reader*. Like the decoder in a transmission circuit, the reader must be precisely matched to the code it is to read: DNA and the cell that reads or decodes DNA into protein molecules must evolve together.

Dennett refers to evolution as an algorithm—a substrate-neutral, mindless, or mechanistic series of steps that will lead in a deterministic (but not necessarily predictable) way to a result. The idea of substrate neutrality (e.g., the series of steps for computing the square root of a number—the result is the same whether the steps are carried out on paper, via a mechanical computer, or by an electronic computer) is linked to the distinction between message and code, form and content, message and meaning. It is also linked to the traditional differentiation between genotype and phenotype, to which I return shortly.

Following up on a suggestion from Dawkins (1989), Dennett proposes that culture is specified by elemental ideas, which he (and Dawkins) calls *memes.* Memes play a role in cultural transmission similar to that played by genes in biological transmission. For Dennett's argument, this close parallel closes the circle through which the fundamental idea of evolution is applied to culture as well as biology:

> Not only all your children and your children's children, but all your brainchildren and your brainchildren's brainchildren must grow from the common stock of Design elements, genes and memes, that have so far been accumulated and conserved... all the achievements of human culture... are themselves artifacts... of the same fundamental process that developed the bacteria.

At a detail level, Dennett's argument is metaphorical; even the word *meme* was chosen to underscore the metaphorical relationship to genes. Yet the process through which these basic units of culture are transmitted and shaped into new cultural artifacts, Dennett claims, is literally the same process as that which governs the transmission of genetic units and their shaping into new organisms. Evolution is an algorithm and works equally well in any substrate.

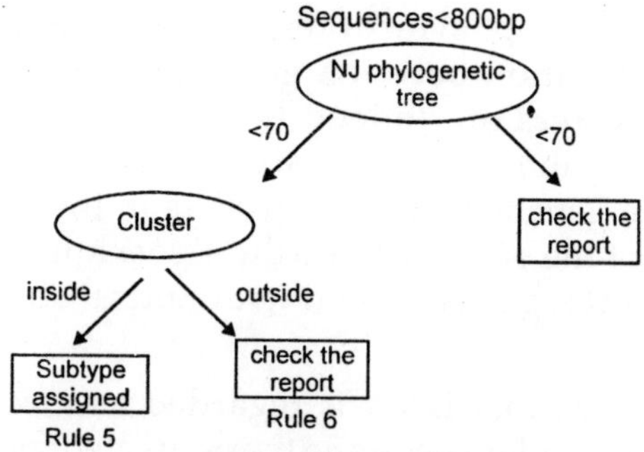

Fig. Genotype

DNA is often referred to as a *genetic code,* and indeed it has several attributes of a digital code. Genes (words) are formed by various combinations of just four nucleic acids (letters). Although this may not be quite as simple as the "dah-dit" of Morse Code, it seems remarkably similar. The similarity goes further: Taken in various combinations, genes specify or contribute to construction of unique proteins. Traditionally, a particular set of genes, referred to as a *genotype,* is distinguished from the organism that is made up of the proteins they encode, the phenotype. It is to this duality that Oyama primarily objects.

The Genetic Code

	U	C	A	G	
U	UUU UUC alanine UUG UUA Leucine	UCU UCC UCA UCG Serine	UAU UAC Tyrosine UAA UAC Stop	UGU UGC Cysteine UGA Stop UGG Tryptophan	U C A G
C	CCU CUC CUA CUG Leucine	CCU CCC CCA CCG Proline	CAU CAC Histidine CAA CAG Glutamine	CGU CGC CGA CGG Arginine	U C A G
A	AUU AUC AUA Isoleucine AUG Methionine	ACU ACC ACA ACG Threonine	AGA AAC Asparagine AAA AAG Lysine	AGU AGC Serine AGA AGG Arginine	U C A G
G	GUU GUC GUA GUG Valine	GCU GCC GCA GCG Alanine	GAU GAC Aspartic acid GAA GAG Glutamic acid	GGU GGC GGA GGG Glycine	U C A G

Fig. Genetic code

However, it is not evident that Dennett's account hinges on a rigid separation between genetypic information and phenotypic expression. Indeed he more or less explicitly incorporates the information in the immediate environment of the cell (the genetic reader) as well as in the more general environment of the organism into the algorithmic process that shapes the intergenerational transmission of genetic information.

To the extent that DNA is regarded as a code and the genome of a particular species as the equivalent of a restricted code (as English is a restricted form of the general alphabetic code and William Shakespeare's style is an even more restricted form of English), the fundamental laws of information theory should apply. The formula for H can certainly specify the maximum information capacity of DNA in general or of any genus' or species' genome in specific. As with the specification of maximum information capacity of human language or of a particular experimental design, whether H is a particularly useful number in this instance is another question.

An obvious application of information theory is suggested by the concept of an evolutionarily stable strategy. The theory of evolution poses two interesting problems for the transmission of genetic information. First, drawing on Dawkins' (1976) metaphor of the *selfish gene* and Dennett's metaphor of an *evolutionarily stable strategy,* it is evident that one of the design problems to be solved by species is the problem of exact replication. A complex of genes that includes coding for accurate, error-free replication, once it had evolved, would remain stable forever.

Moreover, were error-free replication possible, we would expect that it would evolve, eventually, in virtually every evolutionary line. Thus, even if the early adopters of error-free replication were overwhelmed by more vigorous organisms, eventually organisms would evolve that combined a high level of reproductive success within their own ecological niche with error-free replication. The result would be an end to evolution—probably long before multicelled organisms appeared.

The fact that evolution has not ceased requires explanation. Shannon's first theorem provides the required explanation. The first theorem tells us that perfect transmission of information—including genetic information—is impossible. There is no danger that any organism will evolve a way to replicate its own genetic information perfectly generation after generation. It cannot be done. The first theorem also says, however, that clever use of redundancy allows us to come arbitrarily close to perfect, error-free transmission.

Through redundancy (including mechanisms for detecting and correcting transmission errors) it is possible to maintain a high level of genetic stability over a long period of time—as highly successful species, including several species of sharks, have done. The first theorem also tells us where to look for explanations of long-term genetic stability: Look for various forms of encoding redundancy, including mechanisms for detecting and correcting errors as well as mechanisms for preventing errors from occurring in the first place. An obvious form of encoding redundancy is the environment, which acts as a metaphorical editor, deleting alterations in the genotype that lead to unsuitable alterations in the phenotype, and thereby implicitly reinforcing the stability of a successful design.

But there is also a high degree of variation among organisms in the accuracy with which they transmit their genetic information. Some organisms are relatively stable over long periods of time; other organisms (unfortunately including the viruses responsible for influenza and AIDS) evolve rapidly. If it is possible to develop transmission with such a low rate of errors that organisms such as sharks can swim about, effectively unchanged, for millions of years, why haven't other organisms adopted these techniques? The tradeoffs specified by Shannon's second theorem help explain the energy costs involved in pursuing long-term evolutionary stability.

The second theorem tells us that transmission accuracy (and, by implication, long-term genetic stability) comes at a cost. Accurate transmission of genetic information requires a higher channel capacity (more DNA, more energy devoted to preventing and correcting genetic errors), a lower transmission rate (production of fewer offspring), or both. The second theorem also tells us to consider the accuracy/capacity/rate tradeoffs when analyzing the evolutionary strategies of either unusually volatile or unusually stable species.

There seems to be a contradiction here. On the one hand are species like the shark and cheetah, which exhibit long-term genetic stability. On the other hand are species like the viruses responsible for AIDS and influenza, which thrive by virtue of their short-term genetic instability—their ability to evolve new versions faster than the body's immune system can keep up. Therein lies at least part of the answer: At the environmental level relevant to the shark, the ocean changes very slowly. The shark has achieved a high level of perfection as a killing machine, with few natural enemies; any deviation from its basic design would render it less fit.

At the environmental level relevant to a virus, the immune system of a human being (or other animal) changes extremely rapidly: The ability to alter relevant features of the phenotype in a handful of generations, conferred by a highly unstable genotype, is essential to these species' long-term reproductive success. To express this concept in terms of information theory is awkward, if not impossible; it is difficult to see what, if anything would be added to our understanding of these phenomena by formal invocation of information theory.

The information theoretic account poses several other problems, which I discuss in the next section. In brief: Once again, the implicit transmission metaphor needs to be justified. In effect, the transmission metaphor implies that a set of chromosomes is a message encoded in DNA (by the DNA in the parent cells or perhaps by the selective pressures of the environment acting on the DNA) and decoded by the cell.

So closely does DNA resemble a code transmission system, that it is easy to miss that it is not necessarily so. On the other hand, if "DNA As Transmission Circuit" is indeed metaphorical, that does not necessarily imply that it is not a useful metaphor, that its metaphorical entailments are inexact. It does warn us to inspect each of these entailments closely, one by one, before accepting them.

THE CELL AS A DECODER

Oyama criticizes the computational model vigorously, primarily on the grounds that it smuggles dualism back into evolution theory and renders both the cell and surrounding environment as passive recipients of DNA's shaping action. I have focused on Dennett's popularized exposition, with its metaphor of the cell as a "genetic reader," because the problem is most obvious there and because Dennett relies so heavily on the computational model in extending evolution theory to human culture. Oyama's objection seems to be based on the idea that the entire chemical environment within the cell, and indeed the external environment with which the cell interacts, are all part of the unit of evolution.

The "genetic reader" or "genetic decoder" is not a passive Turing Machine, reading the tape of DNA and responding by cranking out whatever proteins are called for. To the contrary, the cell is a complex mixture of chemicals, many of them considerably more reactive than DNA. The only way, Oyama claims, to overcome dualism is to conceive of the process as an ongoing and purely chemical interaction in which DNA plays but one part, often a minor part. As noted in the preceding contrast between sharks and viruses, both stability of the phenotype across generations and many of the variations within a phenotype can be explained by the interaction of cell chemistry with environmental conditions, with no reference to coded messages.

It is helpful to think again about the problem of perfect transmission. For the evolutionarily stable strategy of perfect genetic transmission to have meaning, the environment in which the organism lives and reproduces would also have to be stable. Otherwise as the conditions in which it seeks food and evades enemies change, the organism would become gradually more unfit. From this musing, it becomes apparent that some of the information affecting evolution is in the environment. It is true that perfect genetic transmission is impossible, but that would only become relevant if the environment were perfectly stable over a sufficiently long period. Again the AIDS virus is a perfect example: Part of the reason for its rapid evolution is that the medical profession keeps changing its environment by introducing powerful new drugs and other forms of treatment.

The concept of information is still relevant at some level if only because DNA is so obviously digital. But as with the information capacity of human speech, it may not be useful to have a measure of the information capacity of DNA if other constraints keep the actual transmission rates well below the maxima computed from Shannon's theorems. There has recently been an amusingly dismayed reaction to the news that the human genome may consist of as few as 30,000 to 60,000 genes. However, if the genes interact with each other in pairs, the potential information capacity of even as few as 30,000 genes has an upward limit of 450 million bits.

If the genes interact in larger groups, and if the phenotype is determined by a complex interaction of environment with cell chemistry, including DNA, the upward limit is much, much higher. The information capacity of the genome, as measured by H, may simply not be a useful number, and the constraints suggested by Shannon's theorems may not be the most relevant constraints.

Moreover, Oyama's critique—by dispensing with the separation of DNA from surrounding cell chemistry (as well as from the environment generally) and dispensing with the separation of information from computation— negates the linear model on which information theory, as we ordinarily think of it, is based. There is no source, no receiver, no encoder, no decoder. All that is left is something with some of the surface characteristics of an encoded message. But even this much of the metaphor fails because a genome expressed in DNA does not seem to act like a message expressed in a medium in any conventional sense.

Dennett's exposition, in which DNA is like a computer tape and the cell is a mechanistic reader or transcriber of the tape, implies that DNA determines the entire outcome of the process, with the cell relegated to the role of a transcription machine (decoder) and the environment relegated to the role of selective constraint. Alternatively, one can think of the cell or environment as the active principle, with DNA as constraint. But as Oyama points out, the chemistry of the cell is complex and unitary; DNA is embedded among a complex variety of reactive chemicals. Thus, if Oyama's account is correct, information is not merely in the DNA; it is also in the chemistry of the cell and the surrounding environment. The organism is a product of the entire system, and it is misleading to separate DNA from the rest of it and treat DNA as the *message* and the rest of the cell as the *receiver*.

Oyama seems to reject the traditional phenotype/genotype distinction; it is not clear how seriously this is to be taken. Recent results in cloning experiments certainly undermine a simplistic application of the *code/reader* metaphor. As experience with cloned animals builds up, it has become clear that, despite strong surface resemblances between parent and offspring, the replication is rarely exact. Cloned offspring rarely survive, and the survivors often exhibit deficiencies that undermine their immediate health, their ability to reproduce themselves, or both. Sometimes these deficiencies do not show up until well into adulthood. Humphreys et al. hypothesize that cloning bypasses the normal way in which chromosomes from the two parents interact, and thereby causes some genes to be expressed at abnormal levels.

Not only do genes interact in a complex way with the chemistry of the cell, but DNA from the two parents interacts in subtle and as yet poorly understood ways. Moreover, the chemistry of the cell, the DNA "reader," is itself specified by DNA. All this is a far remove from the long-cherished metaphor of a *genetic code* and the resultant idea of cloning as a process much like printing endless exact copies of an organism with a sort of cellular mimeograph machine. Indeed the concept of a *gene*, which originates in Mendel's pre-DNA breeding experiments, is best regarded as a simplifying metaphor for what is coming to be recognized as a complex set of chemical interactions.

Still evolution does happen, and DNA plays a key role. DNA interacts with the environment, both within the cell and externally, to maintain relative stability of a species from generation to generation. Random changes in DNA interact with changes in the environment to bring about adaptive modifications in the species, eventually, in more extreme circumstances, resulting in the emergence of an entirely new species. *Transmission over a circuit with noise* is a good metaphor for this process and, at some level, is a literal description. To that extent, at least, the theorems of information theory must apply if only to describe the boundaries within which the system must operate.

Information Evolution and Dualism

In electrical engineering, Shannon's theorems have been marvelously productive, allowing engineers to pack a truly incredible amount of signal into a single circuit—as attested to by the high-speed DSL modem that shares a single wire with my telephone. By the same token, Shannon's H and the associated theorems can be used to estimate the maximum capacity of language, of neurons in the human brain, or of any species' genome. The resulting numbers are huge beyond comprehension—a fact that has been celebrated by any number of new-age mystics who use it as a basis for proclaiming the possibility of our every fantasy. However, the real lesson is rather more sobre.

Although as discussed earlier the plot of many a novel or movie turns on an incomplete or garbled transmission, it does not appear that the constraints imposed by channel capacity, transmission rates, and the tradeoff between accuracy and transmission rates play any important role in either the successes or failures of everyday human communication.

All human communication is, in principle, subject to Shannon's theorems, and the energy that fuels the activity of our neurons during communication is in principle subject to the laws of thermodynamics. However, these constraints tell us nothing about how people actually communicate or why they so frequently fail to do so successfully. The theorems associated with information theory have been of some metaphoric use, but the metaphors based on information theory are probably misleading as often as they are insightful.

If I understand Oyama's critique correctly, it appears that a similar point holds with respect to evolution. Given the amount of junk DNA on a typical chromosome, it does not appear that transmission capacity per se is an important constraint on either intergenerational stability or intergenerational change. Junk DNA consists of long sequences of nucleic acids that serve no discernible biological function. Because it also does no harm, it is not selected against and so tends to accumulate over many generations.

If transmission capacity were a constraint on the reproductive fitness of a species, then by definition junk DNA would be harmful (because it would soak up valuable and needed transmission capacity) and would tend to be selected against, hence it would not build up to such a degree. Genetic stability and variety are governed not by the capacity limits of a genetic code, but by the chemistry of the cell and the rigors of the external environment. Dennett's metaphor of the Library of Babel, in which the vast majority of volumes are nonsense, tries to make much the same point. But as it does so, it reintroduces the form/matter dualism to which Oyama so vehemently objects.

Information has a literal meaning with respect to human communication that predates and seems to legitimate Shannon's appropriation of the term. At first glance it appears that Shannon's H simply quantifies this familiar concept, as a number of scholars have argued, beginning with Weaver and Schramm. However, for Shannon's H to work as a literal quantification of the everyday concept of information, it would be necessary to specify a well-defined distribution of alternatives to a given fact, datum, or lore—an objective that proves elusive in practice.

We are left with the conclusion that Shannon's use of the term *information* is a metaphor based on the everyday understanding, a metaphor ironically prefigured by Eliot (1934), "Where is the wisdom we have lost in knowledge? Where is the knowledge we have lost in information?" Communication scholars eager to capitalize on the success of communications engineers in the heady postwar environment of the 1950s turned this metaphor back on itself, making the mathematical concept of information a metaphor for human information in much the same way that the statistical concept of probability has been made a metaphor for epistemological probability.

As I show elsewhere, *probability* began as an epistemological term, distinguishing mere opinion, which was based on reasoning from observed evidence, from true knowledge, which was based on reasoning from first causes or from authoritative sources, such as Aristotle or Holy Scripture. *Probability* referred to the strength of an opinion based on the strength of supporting evidence.

As the science of statistics developed, statisticians linked the distribution of expected outcomes to the idea of probability, at least in part in an attempt to develop a way to quantify arguments in legal cases. Over time the concept of *probability* came to be understood in terms of the distribution of expected outcomes in a chance-governed process. Scientists, especially social scientists, who deal almost exclusively with processes that have indeterminate outcomes, then began to use probability as a description of the strength of an experimental finding.

This epistemological use of probability has been generalized in everyday uses as a metaphor for our expectations of future events even when we have no basis for statistical computations, as for example when we say "the odds are about ten to one that I will miss the party" or, almost interchangeably, "I will *probably* miss the party. " Thus, the concept has come full circle.

Like the *probability* metaphor, there seems to be nothing intrinsically wrong with the *information* metaphor as long as we recognize that we really have no way to apply mathematical information to the task of measuring epistemological information. Yet the two concepts are easy to conflate—and indeed it is difficult to avoid conflating them—precisely because of an underlying literal relationship.

In biology the concept of information can be traced to the Aristotelian theory that male sperm *informs* the neutral medium of the female ovum, and I believe it is this conflation to which Oyama most strongly objects. The classical view of heredity explicitly renders the female as wholly passive and the male as wholly active. Modern biology has long since moved beyond this simplistic and profoundly sexist view of heredity, recognizing that genetic material is contained, in more or less equal measure, in both egg and sperm.

However, the model of DNA as message and cell as receiver/transcriber continues the same basic model in which an active informative principle is separated from a passive medium and ignores the contribution of the information contained in the chemistry of the ovum. The basis for Oyama's whole-hearted rejection of the "DNA As Information" metaphor is that the cell is a highly reactive and complex chemical environment, and DNA is only a small part of this environment. If the role of DNA is conceptualized in terms of information, we also need to recognize the information embodied in the overall chemistry of the cell, not to mention in the surrounding environment. The "Genetics As Information Theory" metaphor seems to work against this recognition because of its connection with the source-message-receiver model.

As is the case with human communication, conflation of metaphoric with literal understanding of genetic processes is encouraged by the underlying literal connections with information theory. DNA is in a real sense a code, and a strand of DNA looks remarkably like a computer data tape. Thus, it is all too easy to make the leap from understanding part of the genetic process as information in a literal sense to understanding the entire process as an instantiation of information theory.

I have previously argued, based on Vervaeke and Kennedy, that it is reasonable and even useful to draw on multiple root metaphors in understanding a complex concept. I have also argued that two major tasks of metaphor analysis are to tease apart the metaphoric from the literal, and to sort out the useful and applicable metaphorical entailments from the misleading and inapplicable entailments. Information theory does apply literally to the genome in some sense at some level. However, if genetic processes are controlled by constraints other than the theoretical maximum transmission capacity of DNA, its literal application may not be particularly useful.

Contra Oyama, I also believe that information theory can be useful as a metaphor to help explain the complex processes of heredity. As Oyama insists, the use of information theory as a metaphor for genetics is tricky and, if poorly handled, must inevitably lead back to the kind of active/passive, form/matter dualisms associated with the Aristotelian theory of heredity. Whether Oyama is correct in her claim that the linkage between information theory and dualism is so strongly habitual in our culture that the *information* metaphor cannot be used without opening the door to dualism remains to be seen: She certainly makes a convincing case for it.

This leads us back to the question of *information theory* as a metaphor for human communication. As I commented in the preceding, there is really nothing wrong with using mathematical information as a metaphor for epistemological information as long as we recognize that we have no way to measure the actual mathematical information content of an epistemologically informative message.

However, extending Oyama's complaint about the information metaphor in genetics, the application of mathematical information theory as a metaphor for the expression of meaning lends itself all too readily to a dualistic separation of meaning from both messages and communicative contexts. Given the history of theorizing about human communication over the past 40 years, one would be justified in extending Oyama's pessimism about the prospects for separating information metaphor from dualism in biology to the prospects for separating the information metaphor from dualism in human communication.

Given the growing evidence that genetic material does not have a simple one-to-one mapping with expressed features of the organism, and that the concept of a gene is probably best viewed as metaphorical, at the very least our understanding of these processes clearly requires some refinement. I have a high degree of admiration for Dennett's ambitious attempt to develop a theory of cultural evolution as an outgrowth of the theory of biological evolution, but it seems increasingly clear that he (and others with similar interests) run the risk of getting caught up in their own metaphors. Dennett has elsewhere made it clear that he does not believe in the kind of mind/body or form/matter dualism to which Oyama objects so strenuously, yet his use and expansion of the information metaphor seems to implicate his ideas in precisely that kind of dualism.

It would seem to be a responsibility of rhetors who use a complex metaphor such as information, as well as of analysts who interpret and evaluate such usage, to do so with careful attention to the entailments of these metaphors. This process must begin with carefully teasing apart the metaphorical from the literal. Once it is clearly understood how a use of information theory is metaphorical, it is necessary to consider carefully each of the entailments: Which of the related metaphors is apt, which is inapt, and which is downright misleading? If one is to draw on and even extend the information metaphor, as Dennett, Dawkins, and many other writers have done, to render a complex concept more comprehensible, there is a considerable responsibility to delimit the entailments and specify the various ways in which the metaphor may mislead.

If one is to extend the information metaphor as a basis for developing the evolution process as a metaphor for cultural processes, as Dennett and Dawkins have both done, it is even more important to be clear as to when the discussion is intended to be taken literally, when it is intended to be taken metaphorically, and, again, where the metaphor may be inappropriate or even misleading.

BIOTECHNOLOGY AND THE ENVIRONMENT

Jeremy Rifkin, president of the Washington, D.C.-based Foundation on Economic Trends, is a futurist who writes about developments in science and technology and their impending influence on economic and social trends. His book *The Biotech Century* explores some of the possible effects on the environment of manipulating nature to produce overnight the kinds of changes that in the past required tens of thousands of years.

While the biotech revolution will reshape the global economy and remake our society, it is likely to have an equally significant impact on the Earth's environment. The new technologies of the Genetic Age allow scientists, corporations, and governments to manipulate the natural world at the most fundamental level—the genetic components that help orchestrate the developmental processes in all forms of life.

In this regard, it is probably not overstating the case to suggest that the growing arsenal of biotechnologies is providing us with powerful new tools to engage in what will surely be the most radical experiment on the Earth's life forms and ecosystems in history. Imagine the wholesale transfer of genes between totally unrelated species and across all biological boundaries—plant, animal and human—creating thousands of novel life forms in a brief moment of evolutionary time.

Then, with clonal propagation, mass-producing countless replicas of these new creations, releasing them into the biosphere to propagate, mutate, proliferate, and migrate, colonizing the land, water, and air. Genetically engineered organisms differ from petrochemical products in several important ways. Because they are alive, genetically engineered organisms are inherently more unpredictable than petrochemicals in the way they interact with other living things in the environment.

Consequently, it is much more difficult to assess all of the potential impacts that a genetically engineered organism might have on the Earth's ecosystems. Genetically engineered products also reproduce. They grow and they migrate. Unlike many petrochemical products, it is difficult to constrain them within a given geographical locale. Once released, it is virtually impossible to recall genetically engineered organisms back to the laboratory, especially those organisms that are microscopic in nature. For all these reasons, genetically engineered organisms may pose far greater long-term potential risks to the environment than petrochemicals.

The risks in releasing novel genetically engineered organisms into the biosphere are similar to those we've encountered in introducing exotic organisms into the North American habitat. Over the past several hundred years, thousands of non-native organisms have been brought to America from other regions of the world. While many of these organisms have adapted to the North American ecosystems without severe dislocations, a small percentage of them have run wild, wreaking havoc on the flora and fauna of the continent.

Gypsy moth, Kudzu vine, Dutch elm disease, chestnut blight, starlings, and Mediterranean fruit flies come easily to mind. Each year the American continent is ravaged by these non-native organisms, with destruction to plant and animal life running into the billions of dollars. Whenever a genetically engineered organism is released, there is always a small chance that it too will run amok because, like nonindigenous species, it has been artificially introduced into a complex environment that has developed a web of highly integrated relationships over long periods of evolutionary history.

Each new synthetic introduction is tantamount to playing ecological roulette. That is, while there is only a small chance of it triggering an environmental explosion, if it does, the consequences could be significant and irreversible. Global life-sciences companies are expected to introduce thousands of new genetically engineered organisms into the environment in the coming century, just as industrial companies introduced thousands of petrochemical products into the environment over the course of the past two centuries.

The reseeding of the planet with a laboratory-conceived second Genesis is likely to enjoy some enviable short-term market successes, only to ultimately fail at the hands of an unpredictable and noncompliant nature. While the genetic technologies we've invented to recolonize the biology of the planet are formidable, our utter lack of knowledge of the intricate workings of the biosphere were experimenting on poses an even more formidable constraint.

The introduction of new genetic-engineering tools and the opening up of global commerce allow an emerging "life industry" to "reinvent" nature and manage it on a worldwide scale. The new colonization, however, is without a compass. There is no predictive ecology to help guide this journey and likely never will be, as nature is far too alive, complex, and variable to ever be predictably modeled by scientists. We may, in the end, find ourselves lost and cast adrift in this artificial new world we're creating for ourselves in the Biotech Century.

4

Environmental Applications of Biotechnology

USE OF BIOTECHNOLOGY

Recombinant DNA and cell fusion techniques may soon allow man to change food crops from inside out with great speed and precision, shaping life to fit the environment. Iowa farmers are divided in their assessments of biotechnology. For the most part, they enthusiastically endorse developments that are expected to improve production efficiencies. But they generally are critical of aggregate increases in food production made possible by biotechnology, as well as likely effects of biotechnology upon farm structure.

The concept of diminishing marginal returns applied to industrial or agricultural production implies that increases in efficiency eventually flatten out, at least until a new technological innovation is introduced. The agricultural sector has experienced stepwise gains in efficiency during certain periods in its modern development. The principal innovations have included the electrification and mechanization of the farm, the introduction of antibiotics in animal husbandry, the hybridization of plant species, and the application of chemical fertilizers and insecticides. Most of these innovations have long run their course. With established technologies, increases in efficiency are made at the margins or through economies of scale. Now agriculture is poised for the innovations of biotechnology.

Leaders of the major agri-industries are evangelical in their optimism over the possibilities. The president of Monsanto, Earl H. Harbison, Jr., described biotechnology as an "economic engine for future agricultural and industrial growth." Not only has biotechnology been cast as a revolution that will enhance efficiency and growth, and enable the world's food supply to keep pace with the expected increase in population, but unlike its predecessors, it has also been characterized as the revolution that will put humans back in balance with nature. The concept of a "softer revolution" has made environmental optimists out of hard-line industrialists. In the words of Monsanto's Harbison: We can at last find biological solutions to biological problems that mechanization and chemicals cannot solve.

The image presented of biotechnology is that the new knowledge of life processes will enable humans to live in a sustainable fashion where economic and ecological efficiency are optimally met. For this to work, the production sector cannot treat nature as a great and infinite wellspring of resources that provides without being provided for.

Have we finally reached a stage in the evolution of technology where new processes of production can be harmonious with natural processes? This expectation carried in the rhetoric of its promoters set biotechnology apart from other industrial revolutions where the criteria for success almost always explicitly favored economic growth and efficiency over the ideas of sustainable production, biocentrism, and ecological balance between humans and the natural habitat.

Can biotechnology achieve such goals? Does this revolution embody the seeds of a gentler technology? Coming as they did in the wake of a new global environmental awareness, the possibilities of releasing genetically engineered organisms into the environment were met with high hopes, skepticism on issues of ecological safety, and some ideological opposition that questioned the wisdom of genetically redesigning plants and animals to increase the human harvest of nature.

THE BACONIAN LEGACY

The desire to control and rearrange species in nature is deeply rooted in the Western scientific tradition. Centuries before genes were understood, scientists contemplated a future in which the natural species were merely starting materials for the conscious rearrangement of the biotic world. For these scientists, the architecture of living forms was likened to the natural minerals of the earth. They were there to be cast into whatever form the human imagination could muster and the life sciences could render.

Francis Bacon, the English scientist-philosopher, unleashed his imagination in a 1622 work titled *The New Atlantis* that describes a biological utopia. Bacon's writings heralded a new philosophy of nature and a new epistemology of science. To have knowledge about the world is tantamount to having power over it. Control over nature was both the goal and the test of knowledge. Bacon provided a literary vision of how scientific knowledge will transform agriculture.

> We have also large and various orchards and gardens, wherein we do not so much respect beauty as variety of ground and soil, proper for divers trees and herbs, and some very spacious, where trees and berries are set, whereas we made divers kinds of drinks, besides the vineyards. In these we practice likewise all conclusions of grafting and inoculating, as well of wild trees as fruit-trees, which produce many effects. And we make by art, in the same orchards and gardens, trees and flowers, to come earlier and later in their seasons, and to come up and bear more speedily than by their natural course they do.
>
> We make them also by art greater much than their nature, and their fruit greater and sweeter, and of differing taste, smell, colour, and figure from their nature. And many of them we so order, as that they become of medicinal use. We have also means to make divers plants rise by mixtures of earths without seeds, and likewise to make divers new plants, differing from the vulgar, and to make one tree or plant turn into another.

Bacon's "biotopia" included an experimental programme for redesigning animals. He described great parks with beasts and birds of all varieties. They were the natural substrate for altering phenotypes to meet human intellectual, practical, or aesthetic needs. By art likewise, we make them [plants] greater or smaller than their kind is, and contrariwise dwarf them and stay their growth; we make them more fruitful and bearing than their kind is, and contrariwise barren and not generative. Also, we make them differ in colour, shape, activity, many ways.

We find means to make commixtures and copulations of divers kinds. We make a number of kinds of serpents, worms, flies, fishes, of putrefaction, whereof some are advanced (in effect) to be perfect creatures, like beasts or birds, and have sexes, and do propagate. Neither do we this by chance, but we know beforehand of what matter and commixture what kind of those creatures will arise. Bacon's utopian prophecy is remarkable to the extent that it operated on the same paradigm that exists in the scientific community today, namely, that there are no unalterable properties of animals or plants. He rejected the idea of species barriers or the notion of integrated wholes. Furthermore, he recognized no moral considerations in the remaking of life-forms. The new secular science was not burdened with the ecclesiastical sanction that humans should not seek to improve upon God's creatures.

THE CREATION OF LIFE

In the latter part of the nineteenth century, German-born physiologist Jacques Loeb advanced the field of biological engineering and believed its fundamental task was to produce living matter artificially. Like Bacon, Loeb refused to accept species boundaries or evolutionary constraints. He actively sought to transcend natural barriers if for nothing more than the challenge of it.

Loeb first received international acclaim for his experiments in circumventing biological fertilization. In 1899, while working with sea urchins, Loeb discovered what he called "artificial parthenogenesis." He devised a method of fertilizing the eggs of sea urchins without male sperm by using inorganic salts. In Philip Pauly's *scientific biography of Loeb*, he notes that "artificial parthenogenesis was a vindication of Loeb's hopes and a model for science to come in which biologists would consciously work to reconstruct the natural order to make it more rational, efficient, and responsive to the ongoing development of engineering science."

Loeb demonstrated that what was believed to be a pure biological function -fertilizing an egg — was reducible to physical chemistry. His discovery brought banner headlines such as "Science Nears the Secret of Life," as well as a host of satirical cartoons. Like Bacon, Loeb rejected the Greek view of science as wisdom and understanding of the world as given. In its place he substituted conquest and transformation as the goals of science, "utilizing the forces of nature to bring about new combinations, creating things which have never been created outside of nature's workshop." For example, Loeb studied the possibilities of redesigning animals that regenerate certain body parts.

> I have succeeded in finding animals in which it is possible to produce at desire a head in place of a foot at the aboral end, without injuring the vitality of the animal. I now have animals that have heads on each end of their bodies; I thus have animals with bilateral symmetry that in nature have different oral and aboral poles. This idea is now hovering before me that man himself can act as creator even in living nature, forming it eventually according to his will.

Loeb developed a specialized vocabulary to describe his research programme in biological engineering. "Abiogenesis" meant the creation of living things from simple substances, whereas "heteromorphosis" described how a physical process could be used to create biological mutants (two-headed tubularia). Through a series of experiments on lower animals, Loeb sought a "technology of living substance".

SCIENCE AND THE GARDEN OF EDEN

After the late nineteenth century, practical efforts to fulfill Bacon's utopian prophecy were initiated by plant breeders who sought perfection first through the process of breeding, screening, and selection, and subsequently through techniques of cross-fertilization and crop hybridization. Plant breeders have altered the genetic strains of crops and flowers by methods developed over centuries, and hybrids have been disseminated widely in the environment. The genetic architecture of the hybrid plant is modified by human intervention, but not by human design, since there is considerable randomness and uncertainty to the process. Millions of plants may be screened in search for a few favorable genotypes.

Modern genetics goes a step further than its classical counterpart. No longer is the plant breeder constrained by species boundaries. As Bacon had imagined, plant species may be transformed according to some archetype or blueprint. Seeds, once thought to be the basic units of plants, are displaced from their privileged status by the cellular and subcellular units.

A similar change has taken place with respect to animal life. The breeding of animals, which can be traced back to the earliest period of domestication, contributes an artificial form of environmental pressure. Evolution is guided by human intention. Thus a sluggish, overweight sow is the desired phenotype for the feedlot, but not, perhaps, for survival in the wild. Industrial and biological evolution became one and the same. The results are animals that are selected for compatibility with humancentered systems — domestic, agricultural, or food processing.

As a method to direct the evolution of animals, breeding is a slow, limited, and ponderous technology in comparison to what may be achieved by the application of molecular genetics to animal reproduction. Since genes can be inserted directly into the fertilized egg, obstacles of species incompatibility are easily overcome. This was dramatized by several experiments on rodents.

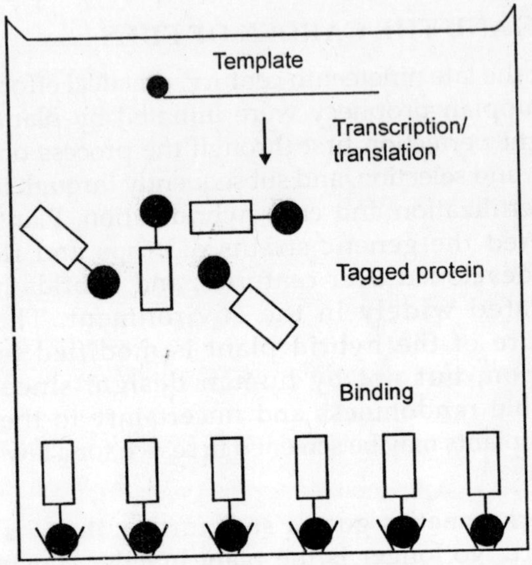

Fig. Rabbit global gene

The first transgenic experiment on an animal embryo took place in 1981 when a rabbit globin gene was inserted into a mouse. In 1982 rat genes and subsequently human genes were implanted into the fertilized eggs of mice. The result was a mouse twice its normal size. Also, with the methodology of germline manipulation of animals it is possible to create identical clones either by splitting the early embryo or by nucleus transplantation.

In the latter case the desired full complement of genes is inserted into a fertilized, denucleated egg. These and other applications of cell mechanics and genetic technology contributed to the commercialization of a new class of environmental products. The following section provides a framework for classifying these research programmes in biotechnology as a prelude to an examination of early public policy responses.

Genetically Engineered Life-forms in the Environment

The idea of controlling and improving upon nature has been a persistent theme in the history of technology. It has always been met with its skeptics, who are no less zealous in their opposition to the hubris of science as the proponents of change are to their attack on nature worship. At one end of the continuum are those that argue there is nothing in nature that cannot be improved upon. At the other extreme are those who maintain that evolution, without human intervention, results in the optimality of life processes. Nature knows best, according to the popular aphorism.

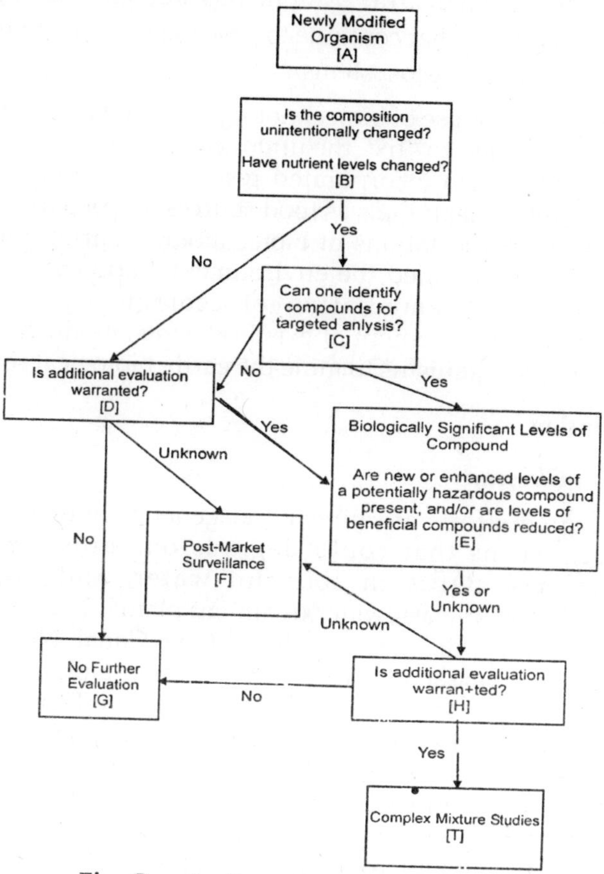

Fig. Genetically engineered life forms

When molecular genetics gave birth to gentechnics, there were already numerous examples of efforts to control and harness nature that had resulted in mixed outcomes. The mining and burning of coal and oil has resulted in new forms of pollution and has been a major contributor to the "greenhouse" effect or the warming of the earth. Synthetic, nonbiodegradable chemicals, which were responsible for the second American industrial revolution, have become ubiquitous residents in the landscape posing remediation problems of enormous magnitude nitude and economic burden. The agricultural system has become hostage to chemical pesticides that continue to pose risks to human health and disrupt natural ecosystems.

The early promise of biotechnology included products and processes designed to "fix" the undesirable outcomes of other technologies. It also incorporated research to improve upon existing plants and animals as food sources. In preparation for reviewing the expectations of biotechnology during the first decade, I have classified the environmental applications into five categories: remediation of ecologically harmful technologies; improvement of existing plants; modification of animals; microorganisms to mine the earth; biological sensors.

Remediation

Pollution Cleanup

Research efforts are directed at genetically engineering microorganisms that could degrade oil spills, detoxify chemical pollutants in soil and water, and improve biodegradation in wastewater treatment plants. The problems of off-shore oil spills, toxic wastes at Love Canal, New York, and Woburn, Massachusetts, polychlorinated biphenyl (PCB) contamination of the Hudson River, dioxin in Times Beach, Missouri, and scores of other cases highlighted the importance of dealing with toxic chemicals in the environment. With the tools of classical microbiology, plasmid engineering, and recombinant DNA techniques at their disposal, scientists began experimenting with microorganisms that break down complex hydrocarbons.

Early studies showed promise of a bacterium that could digest crude oil. In 1980 General Electric won a patent claim for an oil-degrading strain of Pseudomonas, the first such claim issued for a microorganism *sui generis*. Nine years later, the B. F. Goodrich Co. received a patent for a novel plasmid that degrades ethylene dichloride, a compound found in many toxic waste dumps.

Other investigations sought microbes that could neutralize the synthetic herbicide 2,4,5,T (an ingredient in Agent Orange) and PCBs, as well as organisms that could degrade explosive substances and soil contaminated with nuclear materials. The disposal of radioactive wastes has been a persistent problem for the nuclear industry and the Department of Energy. Community resistance to the siting of low-level radioactive waste land fills remains high. The prospect of a biotechnological solution to radionuclides could soften the public's opposition to nuclear power.

Biological Alternatives to Agricultural Chemicals

Promotional materials emanating from the new biotechnology industry spoke enthusiastically of replacing chemical pesticides with biological alternatives that are species-specific and safe for humans and other living things. Some of the new firms embraced the rhetoric of the environmental movement in describing their research agenda and their concern for things ecological. The following is an excerpt from an annual report of the Mycogen Corporation San Diego, California, circa 1988.

> Pesticides are necessary to grow crops economically and preserve the quality of our food supply. Worldwide, a staggering $14 billion are spent annually on crop protection products — $5 billion in the U.S. alone. Despite these huge expenditures, many destructive pests go unchecked and cause major crop losses. The pesticide industry is currently dominated by manufacturers of synthetic chemicals. However, the interplay of environmental and economic forces is creating tremendous pressure for new products which are effective and economical while being safe to use and being compatible with the environment. Biological products are ideally suited to fill the market needs.

Among the major chemical companies, Monsanto was recognized for its strong development effort in agricultural biotechnology. Monsanto's R&D programme included the development of a tomato plant genetically engineered with a protein from the naturally occurring bacterium Bacillus thuringiensis that is supposed to resist insects. Several environmental groups, including the Environmental Defence Fund, the National Audubon Society, and the National Wildlife Federation supported a project of the firm Crop Genetics that involved injecting a microbe called Cxc into the seedlings of corn. This microbe, which ordinarily lives in corn plants, was endowed with a gene from Bacillus thuringiensis, the protein product of which is toxic to certain corn pests.

Another group of agricultural research programmes was directed at slowing down the escalating use of artificial chemical fertilizers, which run off into surface waters and foul lakes and streams by the buildup of organic matter — a process known as eutrophication. The expectation was that plants or symbiont microrganisms could be genetically engineered to fix their own nitrogen from the vast reservoir available in the atmosphere. Agracetus began marketing a microbial soybean seed treatment — a product with genetically modified strains of *Bradyrhizobium japonicum* to improve nitrogen-fixing ability. Another strategy involved transferring expressible nitrogen-fixation (nif) genes of organisms like *Klebsiella pneumoniae* into the chloroplasts of plants, endowing the latter with a selffertilizing capacity.

Agricultural Harvesting and Distribution

Fruits and vegetables are frequently harvested in a preripened stage to accommodate harvesting machines and mass distribution networks. However, the quality of the produce is often affected by being cut off from critical enzymatic changes that take place during the natural ripening process: "While today's tomatoes are the most impressive since the dawn of produce in their resilience, disease-resistance, and travel-hardiness, they are also the most inedible."

Some genetic engineering research is directed at modifying crops so that they will be compatible with the harvesting technologies while also retaining the taste and quality of naturally ripened produce. The challenge for biotechnology is to recapture the taste of produce as it was when grower and consumer were in close proximity, and to accomplish this without disturbing the existing agricultural system and by a means that does not compromise the environment.

Improvement of Plants

Adaptability to Harsh Environmental Conditions

Agricultural crops are adversely affected by many environmental factors, including frost, wind, mineral toxicity, soil acidity, salinity, and of course insect pests. A number of research programmes were devoted to redesigning plants to be more adaptable to harsh environmental conditions. Kenneth A. Barton and Winston J. Brill saw the possibility of mixing and matching in nature by transferring the stress resistant qualities of weeds to food crops. Among the harvests expected of the technology are disease and drought-resistant strains, self-fertilizing seeds, crops with tolerance to high salinity, and plants with enhanced photosynthetic qualities.

Nutritional Value

Improvements sought for existing crops went beyond their growing characteristics and their resistance to natural enemies. Research was also directed at modifying the protein structure of the crops by altering the amino acid chain in the plant's DNA. Among the expected benefits of this research is the enhancement of the usable protein in staple crops like corn and wheat. In their 1983 article in Science, Barton and Brill outlined the possibilities.

The seeds of legumes and cereal grains provide humans directly with approximately 70 percent of their dietary protein requirement. Throughout seed development, storage proteins are synthesized and accumulated within the seed, apparently to provide a source of amino acid reserves during early seed germination. High levels of such protein in seeds provide an enriched amino acid source for both human and animal consumption. However, various deficiencies of seeds in certain essential amino acids do not allow either cereal grains or legumes to provide a balanced diet without supplementation of the limiting amino acids from other sources.

Thus, one approach toward the problem of malnutrition is the improvement of nature's "nutritionally deficient seeds" by encoding them with useful storage proteins. Molecular genetics is viewed as a tool for creating perfect or nearly perfect food sources — foods that carry a complete nutritional balance of amino acids. For lesser developed countries that must import food products to round out the nutritional complement of indigenous crops, or for nations where there has been a scarcity of certain grains, this could be a great asset. But would this technology empower the lesser developed countries? Or would it create a new form of domination by the industrialized nations — a kind of genetic colonialism?

Herbicide Resistance

Nature did not endow most plants with a resistance to synthetic herbicides. Recently, with the extravagant use of herbicides in agriculture and in war as defoliants, plants have been discovered with natural resistance to herbicides. In 1982, herbicides represented 47 percent of the $3.7 billion a year U.S. pesticide market, eclipsing insecticides and fungicides. Biotechnology offered the herbicide industry a way to expand markets and at the same time create a growing environment that simulated a weed-free green house. By engineering crops with herbicide-resistant traits, the herbicide could be used at any stage in the growing cycle.

This so-called improvement upon nature does not take into consideration the effects of the increased use of the herbicide. Guido Ruivenkamp, a Dutch researcher at the University of Amsterdam, wrote that "big industry may see the early opportunity of developing *pesticide-resistant plants* rather than undertaking the longer term effort of developing *pest-resistant plants."*

Plants that Grow Faster

A report issued by the National Academy of Sciences titled Biotechnology: *An Industry Comes of Age describes an enzyme* with the acronym Rubisco that is an important catalyst in photosynthesis. If scientists learn how to move the genes for Rubisco it may be possible to amplify the genes' function and accelerate plant growth. An obvious advantage to the farmer would be the addition of another growth cycle, making more efficient use of agricultural land.

Modification of Animals

Transgenics is the term used to describe the application of rDNA techniques to genetically modify animal species. In some instances, the current research strategy represents an extension of traditional goals to create animals that meet certain requirements of the consumer (as interpreted by agribusiness) or of the food processing industry. For example, there is an interest in creating livestock with leaner meat, and in producing breastier fowl. Since these animals are used exclusively in food production they are not suited to survive anywhere beyond the feedlot. As such they are unlikely to pose environmental problems from a release.

The genetic redesign of animal species has other applications which might affect the environment should the animals be released. There are research programmes to use transgenic animals as functioning, living bioreactors. For example, an animal may be genetically engineered to secrete a human protein in its milk or blood stream. In 1987 a transgenic mouse was produced that secreted a clot-dissolving drug in its milk. Mice are produced twice their normal size by the endowment of a human growth gene.

In yet another case, the gene that codes for the AIDS virus was implanted in the fertilized egg of a mouse and brought to term. As a result, the AIDS gene is found (albeit not necessarily expressed) in every cell of the mouse. Release of such a mouse into the environment could represent a new route for the spread of the disease. It is for this reason that NIH took special precautions in handling these experimental mice by keeping them in high-containment facilities. There has also been experimentation with transgenic fish to determine if they can be genetically designed to survive in colder water or if a larger species can be introduced to repopulate a river. The lingering question is: what impact would such releases have on the aquatic ecology?

Mining of the Earth

Oil Recovery

Microorganisms can be used to enhance recovery methods for removing crude oil from wells. Initial research suggested that certain organisms can reduce oil viscosity, improve rock porosity and permeability, and decrease water-oil interfacial tension. The technique of Microbiologically Enhanced Oil Recovery (MEOR) was first developed in the 1940s. Through genetic engineering, there was a new impetus to improve MEOR by enhancing the traits of organisms like *Clostridium acetobatylicum,* which has been used in the extraction of oil.

The microorganism *Xanthomonas campestris,* which produces xanthum gum, is used for oil recovery by increasing the viscosity of water that is pumped into wells to displace the oil. Another MEOR approach that claims success in recovering oil from marginal wells creates a mixture of microbes and molasses which is injected into oil wells. The gas produced by the microbial growth increases the pressure, forcing out the oil.

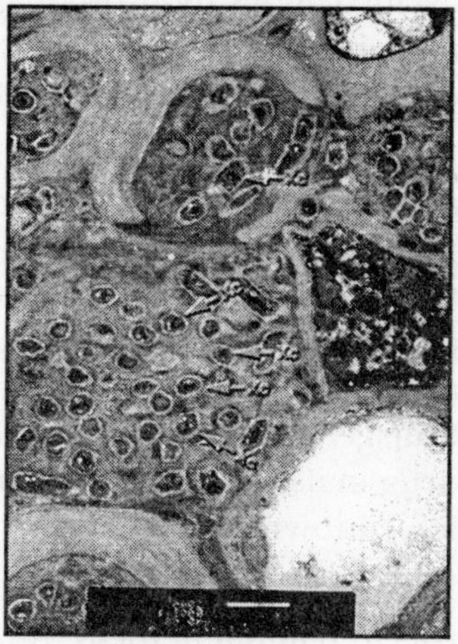

Fig. Xanthomonas campestris

Microbial Mining

Microorganisms have also been useful in recovering certain metals from ore-bearing minerals. Copper and uranium are mined and commercially purified by a process that includes leaching by bacteria such as *Thiobacillus ferroxidans.* The leaching process may be done in situ or in a controlled environment. Efforts are underway to improve the leaching properties of the organisms through recombinant DNA techniques.

- R. H. Zaugg and J. R. Swarz cite some potential hazards in using genetically modified organisms in the mining industry. All foreseeable applications of biological processes involve microbial systems operating in relatively open environments, such as slag heaps or tailing ponds. Consequently, there are risks that microorganisms or their metabolic products will inadvertently contaminate the local ecology.

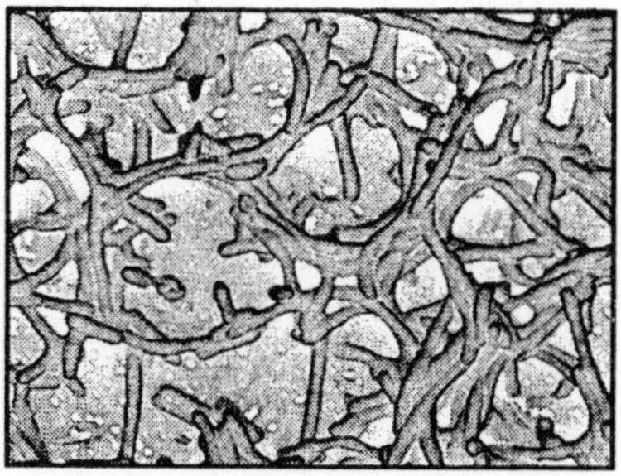

Fig. Thiobacillus ferroxidans

The authors offer three examples:

1. Bacterial leaching operations that generate large quantities of sulfuric acid could acidify water supplies;
2. *Thiobacilli* and related species may acquire the traits to infect humans;
3. Metals concentrated by bacteria from dilute mine waters can accumulate in the food chain.

Biosensors

Microorganisms may also be used as markers in the environment. The concerns about releasing novel organisms have produced an interest in the methods of monitoring the dispersal of the agents. Monsanto Company has developed a genetically engineered strain of *Pseudomonas* carrying natural resistance to antibiotics and genes from *E. coli* that code for proteins that enable the microbe to metabolize lactose. These traits make the organism a possible marker for field tests. The first approval by the EPA of a field test of a genetically engineered organism under the Toxic Substances Control Act (TSCA) was for Monsanto's genetically marked microbe. The test was performed in Blacksville, South Carolina.

The environmental uses of applied genetics constitute an important part of the industrial agenda. But the prospects of releasing genetically novel plants, microbes, and animals have activated a lively public policy debate. The next generation of environmental experiments and applications of biotechnics brought a new group of scientist-stakeholders into the discourse of risks and benefits of new technologies.

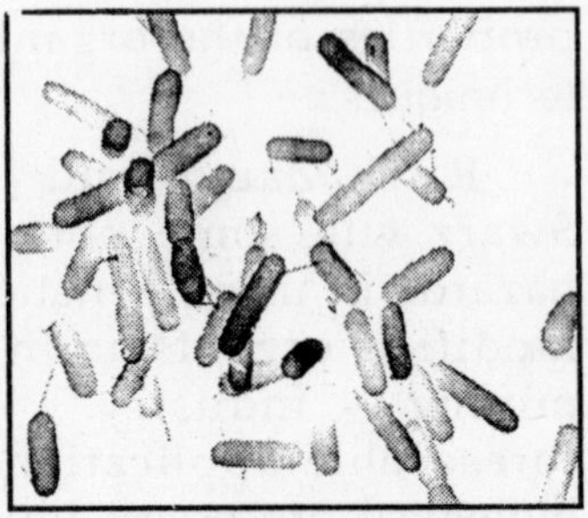

Fig. Pseudomonas

Evolving Policy

The weighty publicity over biotech products and research directions helped bring about a new configuration of public advocacy. Transgenic animals was an issue that linked animal rights organizations, environmentalists, and alternative agriculture groups. The development of more refined genetic screening techniques brought warnings from civil liberties and disability rights advocates. Disclosures of rising expenditures in the Department of Defense's biological defence programme stimulated interest from activists in the disarmament community.

Feminist-health advocacy groups began exploring the impacts of genetic techniques on reproductive technologies. Bioethicists, clinicians, and religious leaders began tackling the thorny problems of human gene therapy. The prospect of major pharmaceutical advances through rDNA research provided the grist for debates in the international health community on the priorities for developing vaccines. Food and agricultural organizations questioned the impact of biotechnology and the new patent provisions on control over plant genetic resources.

It appeared that every major industrial innovation in applied genetics tapped a wellspring of new issues that were brought to the social agenda. Of these, the issue that ignited the strongest public reaction during the early stages of the biotechnology revolution was the introduction of genetically engineered organisms (GEOs) into the environment. This chapter examines the origins of regulatory oversight over deliberate releases and the federal efforts at creating an orderly transition from small-scale laboratory applications of gene-splicing to large-scale releases of GEOs into the environment.

Societal Concern Over Geos

Why has there been so much concern over the risks of releasing plants and microorganisms that have been modified by genetic engineering techniques? Shouldn't the emphasis be on the product and not the particular way the product is created? There are three plausible explanations for the cultural selection of genetically engineered products as a special area of concern. First, the perception of risk associated with deliberate release has largely been formed from prior concerns about recombinant DNA research. In other words, the environmental problems of genetically modified organisms were inherited from earlier stages of the genetics debates. Recombinant DNA-produced organisms are what R. E. Kasperson calls a "social amplifier" in the public's perception of risk.

A second explanation is based upon the notion that genetic engineering provides far greater specificity and control over the product than one could achieve by plant hybridization or breeding of animals. As a result of the specificity of rDNA techniques and their capability of joining quite distant forms of life, the novel life-forms span wider species boundaries and are subject to fewer natural constraints.

By inserting a single foreign gene, a phenotypic property of a bacterium may be radically altered. Resistance to antibiotics is such a property. The ability to change phenotypes in the laboratory with such ease has heightened concerns about deliberate release. Will nature have an opportunity to accommodate to these sudden changes?

To the contrary, some scientists have argued that the specificity of genetic modification makes modern gene-splicing safer than conventional genetics, where genes get mixed randomly and in large clusters. Following this line of reasoning, the precision of gene-splicing means that the resulting properties of the organism will be easier to predict. According to plant geneticist Winston Brill writing in *Science,* in conventional breeding it is impossible to predict the properties of the progeny from most of the crosses. Genetically engineered plants have greater specificity: "If we compare plants derived from breeding programmes with those derived through genetic engineering, it is clear that, in the latter case, the addition of a few characterized genes to the plant results in properties that are relatively easy to predict."

From another perspective, despite the specificity of rDNA techniques, with them one might be capable of producing more substantial changes in organisms with fewer genetic alterations than with classical genetic techniques. The issue of whether modern genetic engineering techniques are capable of producing varieties of plants, microbes, and animals that could never have arisen from the natural rearrangement of genes remains unresolved. It is widely acknowledged that we humans can at least accelerate or redirect the evolutionary process even if we cannot create qualitatively new life forms.

A third reason why genetically engineered plants and organisms designed for environmental release have attracted more concern than the release of similar products prepared through conventional genetics is related to the reputed power of the new technology. Gene-splicing has been the raison d'être of a technological revolution. This is not simply another discovery in the slow, incremental growth of science. This discovery has given birth to a new industrial process for radically reconfiguring biological matter.

The disclosure that there is a new power to transform nature is one of the sources of public and scientific anxiety. It might be argued that if rDNA technology embodies a power to stimulate the growth of a multibillion dollar industry, why should its risks be considered comparable to those of conventional genetics? What is the likelihood that the industrial potential of gene-splicing (gentechnics), which is, let us say, a thousand times greater than that of conventional biotechnology, will be unleashed without any increases in environmental risks? It is certainly a question worth considering.

Setting aside for the moment the body of scientific argument about the potential risks of new biotechnologies, there is an undisputed equation between technological power and risk anxiety that must be considered in fully understanding the public reaction to biotechnology. The simultaneous pronouncements about power and safety seem incongruous to a popular culture that has been sensitized to technological failure.

In trying to comprehend the risks of releasing genetically altered species into the environment, inevitably we are drawn to comparisons. Two technologies of commensurate transforming quality to gene-splicing are synthetic organic chemistry and nuclear physics. Both of these technologies are capable of creating new arrangements of matter in a fashion analogous to the creation of novel species through biogenetic engineering.

Risk assessment for synthetic chemicals has been in progress for several decades. There have been some important breakthroughs as well as notable impediments. The identification of a chemical substance is a well-defined process. It is, therefore, possible to construct a precise inventory of chemical compounds. The same is not true for biological agents, at least in some practical sense. Microorganisms and plants are classified by phenotype, and therefore the addition or deletion of a few genes will not necessarily warrant a change in the classification.

It is estimated that 60 to 80 thousand distinct chemicals are used in industry out of a pool of several million that have been synthesized. If a genetic identification system was used for biological organisms, the number of extant chemicals would pale against the number of distinct life-forms since, for the latter, a single nucleotide change would be a differentiating factor. Keeping track of novel organisms and establishing an identification system is a problem of enormous complexity, and probably unrealistic since genetic mutation is a constant occurrence. Yet any serious regulatory effort in biotechnology must address the identity question.

An obvious difference between inert chemicals and life-forms is that the latter are self-reproducing. Throughout the history of the chemical industry there have been countless cases where toxic carcinogenic chemicals were disposed of in lagoons and landfills. These chemicals saturate the soil and eventually migrate to subsurface water supplies where they contaminate drinking water. Once embedded in the earth, many industrial chemicals are difficult to remove. Entire neighborhoods in areas such as Times Beach in Missouri and Love Canal in New York have been evacuated because of toxic waste. In other, more manageable situations, contaminated soil is removed or filtration methods are applied to poisoned wells.

The mistaken release of a nuisance biological agent cannot be handled by techniques developed for chemical contamination. At the worst, the released organisms are beyond recall and will grow to population orders of a magnitude beyond the density of the inoculation. Moreover, if a novel organism were introduced and subsequently found to be dangerous, geographical isolation and community evacuation would simply not work.

Considerable progress has been made in standardizing toxicological testing for chemicals. The use of accepted methodologies, standardized target species such as germ-free mice or rats, and microbial assays such as the Ames Test have contributed to uniform standards of risk analysis. Notwithstanding the progress, there are still many areas of uncertainty and scientific debate. Among them are questions about dosage and extrapolation from animal to human effects. Also, human epidemiological studies are frequently too insensitive to pick up small increases in cancer incidence over a lifetime exposure. While there are many effects of chemical exposure that are not well understood, there is at least a basic methodology for gathering the data.

There is no commensurate methodology for assessing the risks of released organisms. Moreover, the risks associated with certain chemical releases are real. Their effects on humans and the biotic world have been observed. In contrast, the potential risks of genetically altered life-forms are currently speculative. As a result, the social demand for evaluating the risks of bioengineered products designed for environmental use may not evoke the same urgency as if the hazards were confirmed.

When chemicals enter the environment, it is not always obvious what effects the breakdown products (metabolites) will have on the ecology. Chlorine has been used extensively to purify drinking water. Its use has been associated with the appearance of chloroform, a potent carcinogen, and other troublesome compounds called tribalomethanes.

While there may be risks in the continued use of chlorine, the alternatives are not good. No safer method for purifying water is available. We can never be sure how released chemicals will reconfigure themselves in ecosystems, but compared to the possibilities of biological entities, the range of unexpected outcomes for inert chemicals is probably much narrower since the biological entities mutate. A subclass of all possible mutations affects the phenotype of the entity.

Whether it is more complicated to predict the mutational possibilities of a novel organism than to predict the synthetic pathways and metabolites of a new chemical is the subject of debate. Organisms are certainly more complex than inert chemical compounds. That does not mean that they are inherently riskier, but it does portend a high level of complexity in analyzing environmental risks of genetically novel entities in contrast to newly synthesized chemicals.

There are also some comparisons to be made between biotechnology and nuclear technology. The number of radionuclides is relatively small, surely in comparison to industrial chemicals. They are generally used in well-defined and controlled settings. There are laws regulating the release of radioactive materials. Also, the health effects of radionuclides in high and moderate doses have been studied and are reasonably well understood. This is not the case, however, for low-dose, long-term exposure.

Radioactive materials are detectible in minute quantities with a sensitive monitoring device. To improve the confidence of residents living in the vicinity of a nuclear power plant, some communities have been provided with radiation detection counters.

There is nothing analogous in biotechnology. In theory, one can identify and track bacteria that have been released into the environment. The organisms would have to be tagged in a special way. Even then the identification can be a difficult task depending upon the behaviour of the microbes and the conditions of the environment. Each case is unique. At this time, there are no standard methods of detection and no cannonical procedures for distinguishing safe from deleterious organisms. In comparison to bacteria, it is much simpler to detect the spread of an unwanted plant (a weed). But once released, plants, like bacteria, may be impossible to recall.

We have seen how, within a period of a decade, a single critical discovery was the progenitor of an industrial revolution. The investment, scientific, and corporate communities moved expeditiously to capitalize on the commercial opportunities of the new genetics. To whom was this new industry accountable? How were the public policy issues handled? What social guidance was imposed upon the new technological direction? The next section discusses the early regulatory response to environmental applications of biotechnology.

Beginning in the 1980s, industry and university proposals for field-testing genetically modified plants and organisms triggered a major science policy debate in the United States that spilled over to the European community. Those who have followed the recombinant DNA controversy from its inception will recognize that the current configuration of policy alternatives is the result of a ten year historical process. Initially, molecular geneticists cast the problem of genetic engineering in technological terms. Gradually, public perception of the problems associated with gene-splicing focused attention on the ethical and ecological issues.

The emergence of a second generation of genetics policy debates brought participation from new disciplines, new communities, new public interest groups, and new federal agencies. Public concerns slowly shifted from the singular issue of laboratory safety to a much broader range of problems. And while these changes were taking place, regulatory oversight of biotechnology also shifted from the National Institutes of Health to other governmental bodies.

For nearly a decade, the agency that assumed primary responsibility for the safe uses of genetic engineering was the National Institutes of Health (NIH). Essentially a science-funding agency under the Department of Health and Human Services, the NIH established the Recombinant DNA Advisory Committee (RAC) in 1974 at the recommendation of Paul Berg, the Stanford University biologist who provided leadership in the early efforts to assess the risks of rDNA research.

According to Berg's plan, an international meeting of biologists would result in a set of principles for safe handling of rDNA molecules. Those principles were then to be used by the NIH's newly formed scientific advisory committee to establish guidelines for all genetic experiments involving the cutting and splicing of foreign genetic material. Berg, along with other scientists who organized the Asilomar meeting, surmised that if the NIH did not act with dispatch in responding to the potential risks of gene-splicing, Congress might pass restrictive legislation.

They viewed the passage of such legislation as detrimental to the interests of biology; particularly at stake was the legitimacy and progress of molecular genetics. First, biologists could lose influence over the risk assessment and risk management process. Second, the field of molecular genetics might become stigmatized as the only discipline whose principal research method was regulated. Third, there was considerable concern that rDNA legislation would be inflexible and difficult to amend. Asilomar organizers feared that biology would be saddled with irrational requirements.

Written by scientists for the general use of scientists, the NIH guidelines made their debut in June 1976. No explicit references were made to industrial processes or non-NIH supported uses of rDNA. The guiding principle behind the development of the guidelines was containment. Since one could not predict at the outset with any reasonable degree of certitude that a foreign gene introduced into an organism could not inadvertently transform it into an epidemic pathogen, the consensus at Asilomar was to construct a set of containment provisions consisting of physical barriers, safety operations, and carefully selected host organisms chosen because they do not survive well outside of the laboratory setting.

Each containment level was matched with a class of experiments that was permissible under the stipulated conditions. As scientists on the RAC and elsewhere became more confident in the safety of rDNA techniques, laboratory containment requirements were substantially relaxed. The early NIH guidelines contained a provision that restricted industrial applications of rDNA technology. Cultures of rDNA organisms produced or handled in volumes greater than ten liters (classified as large-scale) were prohibited for NIH grant recipients.

The large-scale prohibition was not based upon a scientific assessment of risk. It was a convenient threshold introduced by academic scientists to protect the use of standard laboratory beakers in basic research. A second provision of the NIH guidelines restricted industrial activity by explicitly proscribing the intentional release of an rDNA organism into the environment. Since absolute containment was nothing more than an idealization, unintentional releases were considered unavoidable.

However, by limiting the volume of rDNA culture, the probability of escape could be minimized. Also, since the volume of agent released is correlated to survival and propagation, the large-scale prohibition also supported the general containment strategy. Commercial interest in rDNA techniques grew rapidly in the mid- 1970s. Estimates of industry growth vary. My own count indicates that a minimum of 14 new biotechnology enterprises (NBEs) were formed in 1976, the year the NIH guidelines were introduced. During 1979, NBEs grew by at least 26; and in 1981, at least 66 NBEs entered the biotechnology industry.

The NIH had no legal jurisdiction over the research in private companies. In practice, however, firms were not willing to risk the negative publicity that might arise if they violated the NIH guidelines. The nascent biotechnology industry was comprised predominantly of small firms started by relatively young scientists, many of whom retained their university affiliation. The close link between academe and industry may help explain the high degree of compliance among new firms with the standards adopted by the NIH.

Since these scientists were groomed on the NIH guidelines, the complications of compliance that might have beset a new industry were minimized. Despite a watchful media, there is no evidence that the NIH guidelines were flaunted by the biotechnology industry. Ironically, the few cases where violations of the guidelines were reported took place at universities. Geoffrey Karny cited two factors responsible for industry's compliance with the voluntary guidelines: "First, the possibility of tort lawsuits provides monetary inducement to comply with the *Guidelines,* which would probably be accepted as the standard of care against which alleged negligence would be evaluated. Second, the threat of statutory regulations, which the companies have sought to avoid, always exists."

Between 1976 and 1979 the NIH process for overseeing rDNA research was put to its severest test. First, the city of Cambridge, Massachusetts, issued a moratorium on rDNA experiments requiring moderately high physical and biological containment. After a widely publicized citizen review process, the city passed the country's first rDNA law in 1977. The law departed from the NIH guidelines in a few minor respects. More importantly, it symbolized the right of local government to exercise control over where the research gets sited and the safety conditions of its performance. Moreover, it established uniform and legislatively mandated guidelines for both publicly and privately funded research.

After the Cambridge rDNA law was passed, nearly two dozen states and local communities debated the issues. Legislation was enacted in about half the jurisdictions. In response to local events and a national mood of concern toward gene-splicing, fifteen distinct bills were filed in Congress between 1977 and 1978 to regulate rDNA research. Some of these bills would have shifted the regulatory authority from NIH to a national commission. These bills also varied in the degree to which local laws were subject to federal preemption. Congress spent two years debating the issue of an rDNA law. A compromise bill was finally voted out of committee early in 1978, but for lack of strong congressional leadership and interest it failed to reach the House or Senate floor for a vote.

Congressional failure to enact legislation strengthened the NIH's position as the sole agency overseeing rDNA activities. Responding to continuing public concern over the research, the Department of Health, Education, and Welfare (HEW, currently the Department of Health and Human Services) rewrote the RAC's charter and increased the size of the committee from sixteen to twentyfive. The new charter, issued in 1978, stipulated that one third of the committee was to consist of individuals with expertise and interest in public health and the environment.

The change in the composition of the RAC drew sharp criticism from prominent scientists who argued that rationality was being compromised by including nonscientists in what was essentially a technical process. When the expanded RAC met in early 1979, its agenda was filled with petitions for relaxing containment requirements and approving additional host-vector systems.

Over the next few years several important changes were made in the rDNA guidelines that established a role for the NIH in the review of field tests. A voluntary compliance programme was established that gave the private sector access to the NIH review process, while prohibitions against large-scale rDNA activities and the intentional release of genetically altered strains into the environment were removed.

Compliance Programme

After the decision by Congress not to enact rDNA legislation, there was a strong residue of criticism that the NIH guidelines could not protect society from the potential adverse consequences of commercial gene-splicing. In response, NIH developed a voluntary compliance programme for institutions that did not fall under the agency's purview. This initiative gave the fledgling biotechnology industry the opportunity to gain the imprimatur from the NIH for both laboratory and commercial-scale rDNA work.

A firm wishing to participate in the programme first submitted the composition of its institutional biosafety committee (IBC) to the NIH's Office of Recombinant DNA Activities for approval. Once IBC approval came, a firm could file requests with the RAC following procedures similar to those of university petitioners. One difference between the NIH's handling of academic and industry proposals is that, on the occasion of the latter, the RAC went into closed session. Members were required to sign confidentiality pledges for the protection of information deemed proprietary by the firm.

Some RAC members were opposed to having the committee review proposals in closed session. One public interest member refused to participate in the review of industry proposals. He argued that oversight of the private sector was the responsibility of those agencies of government with statutory authority to protect workers, public health, and the environment.

Since the NIH lacked authority to carry out these functions, he felt his participation would give legitimacy to this extension of the NIH's role. Another RAC member expressed the following sentiment: "Voluntary compliance is the worst of all possible worlds. You achieve none of the objectives of regulation and none of the benefits of being unregulated. All you're saying is 'I give a stamp of approval to what I see before me without any authority to do anything.'

The programme was also the target of mainstream critiques. As an example, in its 1981 report on biotechnology, the Office of Technology Assessment wrote: "The most significant limitation in the scope of the Guidelines is their nonapplicability to industrial research or production on other than a voluntary basis. This lack of legal authority raises concerns not only about compliance but also about NIH's ability to implement a voluntary programme effectively."

In May 1979, the NIH's advisory committee went on record opposing voluntary compliance by a vote of nine to six, with six abstentions. The RAC voted that non-NIH funded institutions should be required to comply with the guidelines. This recommendation notwithstanding, the voluntary compliance programme became a permanent part of the guidelines in January 1980.

Many biotechnology companies considered it in their long-term interest to secure RAC approval for their rDNA activities. Regardless of how the firms' management felt about the potential risks of gene-splicing, submitting experiments for NIH approval was excellent public relations. The voluntary compliance programme helped the biotechnology industry respond to the criticism that the private sector was operating without regulations. When pressed by local communities to demonstrate the safety of genetic experiments, a company's most compelling response was that it complied with the NIH guidelines.

In the wake of sporadic episodes of local opposition to genetic engineering research, the biotechnology industry sought a predictable and stable regulatory climate, but one that would not impede research and development. The NIH contributed to this goal, but with limited success. The voluntary compliance programme proved functional to the industry during its early years of development when there was intense competition, investment instability, and the uncertainties of local regulation. As the commercial activities progressed from laboratory research to product development, the NIH policies accommodated to the new stage of industrial activity, particularly in their response to large-scale work and the release of genetically modified plants and organisms into the environment.

Large-scale Prohibitions

Provisions were built into the early NIH guidelines for waiving the large-scale prohibition in cases where minimal risks were balanced against important societal benefits. The wording of the prohibition was clarified in the 1978 version of the NIH guidelines. No exceptions to the prohibition against the production of largescale cultures were permitted "unless the recombinant DNAs are rigorously characterized and the absence of harmful sequences established." Under NIH's leadership the risk assessment paradigm was in large measure still under the primary influence of geneticists.

For a short period of time, the NIH gave serious attention to all phases of large-scale work with rDNA molecules. Proposals submitted to the RAC were required to include a description of laboratory practices, specifications on physical and biological containment, risk data, characterization of genes, and the design of the fermentation equipment in conjunction with the physical description of the facility. In 1980, the RAC published a standard that it planned to use for reviewing large-scale proposals.

The committee's role in evaluating plant design and operations for large-scale fermnentation drew criticism from some members. They argued that the RAC should confine itself to advising the NIH on the nature of biological procedures and not plant operations. The committee, after all, had no special expertise in bioprocess safety engineering. It depended on outside consultants for guidance in this area. Also, it was a matter of some significance that molecular biologists on the

RAC found the review of plant design boring and uninformative.

Other RAC members contended that if the voluntary compliance programme was to mean anything, a comprehensive review of large-scale operations was essential. Since no other federal agency was engaging in this review, they believed that it was the NIH's responsibility to fill the regulatory void. The internal debates over this issue were a poignant reminder of the NIH's ambivalence in serving as overseer of commercial genetic engineering.

These debates also cast doubt on the logic of having a biomedical funding agency guide a nascent industry exclusively through a system of voluntary measures. The peculiar nature of this regulatory programme began to reveal its contradictions. For example, within the RAC opposing factions interchanged positions. Initially, the group most skeptical about the safety of genetic engineering expressed opposition to the RAC's review of industrial proposals, particularly large-scale ones.

They reasoned that the NIH was acting as de facto regulator without enforcement powers or congressionally derived authority. If the RAC refused to serve this function, members of this faction believed that agencies like the Occupational Safety and Health Administration (OSHA) and the Environmental Protection Agency (EPA) would enter the field. Concurrently, there were other members of the RAC who opposed broader federal involvement in biotechnology, particularly legislative or rule-making actions. They viewed NIH's continued role in overseeing commercial rDNA work as essential to preempting such initiatives.

The unexpected reversal took place among committee members around 1980. The pro-regulatory group grew less sanguine about the involvement of other agencies in regulating biotechnology. As a consequence, this faction began supporting a stronger role for the RAC. Meanwhile, the committee's regulatory minimalists, also confident that broader agency involvement was unlikely, backed an NIH role over commercial rDNA activities that were limited to issues of pure genetics, namely, sequence characterization and approval of host-vector systems.

The regulatory minimalists succeeded in so limiting the RAC's review of the safety of rDNA products. By 1981, firms taking part in the NIH's voluntary compliance programme were no longer obligated to submit data on sterilization of their fermentation system or their procedures for disposal of rDNA cultures.

This policy change was expressed in a proposition adopted at the September 1980 RAC meeting: "The RAC will determine if a given recombinant DNAcontaining strain is rigorously characterized and the absence of harmful sequences is established. Such a determination shall include specification of a containment level. These determinations should not in any way be construed as RAC certification of safe laboratory procedures for industrial scale-up."

The potential release of rDNA organisms through the effluent of bioreactors did not attract any public attention. While the EPA had jurisdiction over such biological releases under the Clean Water Acts, a constituency for agency action did not develop. The newly formed Committee (later Council) for Responsible Genetics published an article on biogenetic waste in its public interest bulletin *GeneWatch*.

However, the issue of biogenetic waste was not a rallying point among environmentalists or the general public. It seemed to lack some important features that influence risk selection. First, it did not impart a clear and present environmental danger. Second, the waste stream, even if it carried genetically engineered organisms, did not excite the media. Bioeffluent was not intentionally designed to alter nature. Since science writers are captivated by the frontiers of science, there was not much of a story in fermentation sludge.

Quite a different media and public reaction took place when proposals appeared before the RAC requesting approval to field-test genetically modified plants and bacteria. There is an informative distinction to be made in the public reaction between intentional and unintentional releases of genetically modified life-forms. To fully grasp the distinction, we have to look at the kinds of experiments proposed, the types of media coverage they were given, the role of scientists in raising safety concerns, and the development of a public interest constituency.

Deliberate Release of GEMs

As the RAC reduced its oversight over industrial bioprocesses, it became more active in reviewing environmental releases of genetically engineered microorganisms (GEMs). Consequently, field-testing was given greater visibility in the media. From 1980 to 1984, when companies and university scientists were preparing to field-test their products, the NIH was the only federal agency with active responsibility over deliberate release experiments involving genetically altered plants or microorganisms.

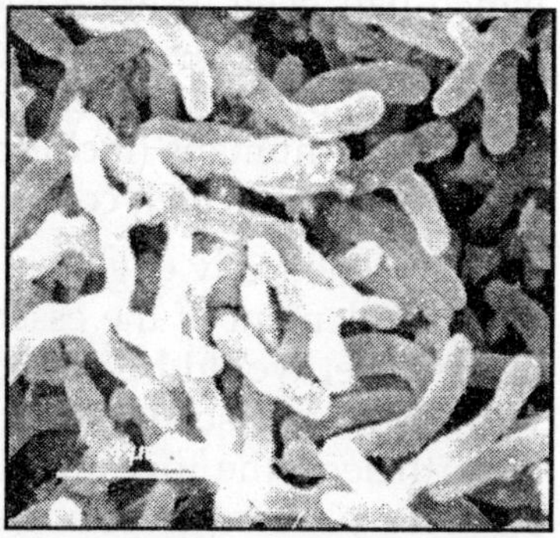

Fig. Genetically engineered microorganism

Progressive relaxation of the rDNA rules finally led to the removal of barriers to field-testing. The first revisions of the guidelines in 1978 still prohibited "deliberate release into the environment of any organism containing recombinant DNA," but individual waivers were permitted after proper public notification, RAC review, and approval by the director of the NIH. The revised NIH guidelines of 1982 eliminated the entire list of proscribed experiments. By June 1983, the prohibition against intentional release of rDNA organisms was replaced by a multitiered review process. Submissions for deliberate release required approval by the RAC, the institution's biosafety committee (IBC) and the NIH director, in addition to various subcommittees.

Agricultural applications of biotechnology, widely publicized in the media, were being readied for field trials. By December 1983, the RAC had reviewed and approved three proposals for releasing genetically altered life-forms. In each case, the committee concluded that the tests posed no significant risk to health or the environment.

The first three proposals for deliberate release came from university scientists and therefore did not involve proprietary information. In March 1980, a Stanford University scientist requested approval from the RAC to field-test a corn plant into which had been inserted the corn storage protein gene with modified sequences. The genetically altered strain included a new corn protein gene that encoded all the amino acids essential to humans (including an enhancement of lysine and methionine, in which corn is deficient).

By genetically engineering the new corn protein with the full complement of the essential amino acids, its value to the human diet would be improved. The RAC failed to come up with a hazard scenario for the genetically modified corn; nevertheless, the investigators were required to detassel the plant to prevent pollen dispersion during the field trials. Permission for the test was granted in August 1981. There was no significant public reaction to the decision.

A second proposal for field-testing was brought to the RAC by a Cornell University scientist in June 1982. Tomato and tobacco plants were transformed with DNA from yeast and *E. coli* to provide them with antibiotic resistance. In reviewing the experiment, the RAC raised concerns about the possible spread of antibiotic resistance and the effect of an itinerant plant on the ecology. Neither of these concerns delayed the committee's approval, which it gave on October 25, 1982. When the recommendation reached the director of the NIH, he referred the proposal to the U.S. Department of Agriculture (USDA) for a reading. After USDA approval was given in February 1983, NIH accepted the field test in April 1983, nearly a year after the proposal was first brought to the RAC.

The third of the early field test proposals for genetically modified life-forms proved to be the most controversial. Ironically, it was viewed by some experts as among the safest deliberate release experiments one could perform in the environment. There was one important distinction between the first two proposals and the third: the latter consisted of genetically altering a soil bacterium.

The difference between plants and bacteria proved to be a critical factor in the public perception of risk. Also, there was an established tradition of introducing hybridized plants into the environment. However, no analogous tradition existed for microorganisms. The relative novelty of this field test was reflected in the RAC's review.

In September 1982, scientists at the University of California at Berkeley proposed to field-test two soil organisms (*Pseudomonas syringae* and *Erwinia herbicola*) from which about a thousand base pairs of DNA sequences had been deleted. In their natural state these organisms synthesize a protein that provides a nucleation point for ice crystallization, and are known as ice-nucleating agents (INA). By excising the genes responsible for ice nucleation and establishing the genetically modified organism in the environment, scientists believed they could reduce the temperature at which frost begins to form on crops.

When the proposal for field-testing the microbes with the deleted ice-crystallization gene (denoted INA$^-$ or ice minus) was first brought before the RAC, one member expressed concern about the role the organisms might play in the environment. Questions raised about the effect of INA$^-$ on precipitation patterns were based upon some theories that INA$^+$ (normal ice-nucleating agents) played a role in atmospheric weather. The precipitation issue was not resolved by the committee.

Nevertheless, the RAC approved the field test, although by a small plurality (seven yes; five no; two abstentions). When the recommendation was transmitted to the director of the NIH, more data along with additional safeguards were requested. Several months later on the second round of review, the RAC approved the proposal unanimously. The NIH gave the final go-ahead for the field test in June 1983, some nine months after it had initially received the proposal.

Investigators agreed to mark the strains of the ice minus organism with antibiotic resistance, which allowed them to monitor the dispersal of the organism. They also agreed to limit field-testing to a single location, the University of California agricultural field station at Tulelake in northern California, a site isolated from the major fruit tree and citrus-growing regions of the state. Despite these precautions, the RAC's decision on the field test for ice minus was met with citizen opposition and litigation from the Foundation on Economic

Trends, Jeremy Rifkin's organization based in Washington, D.C. Between 1980 and 1985 the RAC reviewed five proposals for field-testing GEMs. Four were approved and one was voluntarily withdrawn.

Genetics and the Environmentalists

In the mid- 1970s, when rDNA molecule research was still an exotic technique, the public entry into the debate over its safe uses was determined by several factors. First and foremost, a group of distinguished biologists called attention to the new technique by publishing letters in science journals that caught the attention of the popular press. The international meeting at Asilomar was attended by science writers, whose coverage of the events dramatized the disagreements among biologists at the meeting. In drawing attention to the new discovery, the Asilomar organizers did not intend for the fate of the new technology to be decided by popular acclamation or democratic process.

However, some biologists, frustrated by the NIH's role in setting safety standards, brought their concerns to university campuses and local communities. To the public, it appeared that scientists were polarized on the dangers of rDNA research. Second, the issues were dramatized through the world press by the controversy that erupted in Cambridge, Massachusetts, which pitted community values against the interests of national science. The 1976 Cambridge rDNA controversy represented the birth of public involvement in genetic engineering.

Prior to that, the issues were debated exclusively in professional groups and on university campuses, notably the University of Michigan. Third, national environmental groups such as Friends of the Earth, the Environmental Defence Fund, and the Sierra Club devoted some attention to genetics issues between 1976 and 1979. But the commitment by these groups was limited usually to a single staff person. Their activities were met with skepticism, and in some cases, disapproval by some board members who had ties to the biomedical community.

Fourth, but to a lesser extent, litigation had drawn public attention to the issues. Two suits were filed against HEW/NIH in 1977. A resident of Frederick, Maryland, petitioned the court to enjoin the NIH from performing a risk assessment experiment until an environmental impact statement was prepared. A second suit filed by Friends of the Earth alleged that HEW failed to comply with the requirements of the National Environmental Policy Act when it established policies for rDNA research.

Fifth, congressional interest in the safety of rDNA research intensified in the mid-to-late 1970s. House and Senate hearings provided a forum for activist scientists and environmental lobbyists. The accumulated effect of these factors resulted in heavy media coverage for both local and national events related to gene-splicing experiments.

The subsequent public debate over the specific issue of the deliberate release of genetically engineered organisms has been fostered by a similar configuration of factors, namely, scientist-critics, litigation, environmental coalitions, local opposition to experiments, and congressional interest. However, the old and new rDNA debates differ in the weighting of these factors. For example, Jeremy Rifkin has proven particularly effective in drawing public attention to the environmental and ethical issues associated with the deliberate release of GEOs. Operating with the modest resources of his nonprofit organization, Rifkin has utilized a variety of techniques, including confrontational protests, litigation, and short-term coalition building, to dramatize breaking issues in genetics and public policy.

Rifkin's philosophical opposition to genetic engineering has been expressed in his writings and media accounts that feature his work: "Genetic engineering represents the ultimate negation of nature." On genetics policy, he has had more impact on the media that any single group or individual in the United States. His notable success as a publicist and genetics critic can be explained by several factors. Rifkin can act quickly since he is not accountable to a board of overseers or a mass organization. He is able to identify and capitalize on the weakest link of any policy process.

With properly timed and targeted litigation, Rifkin has been able to strike at the jugular of the bureaucracy. He has proven time and again that a well-placed lawsuit is a magnet for media attention. Rifkin made his national debut as a genetics activist in 1977 when he and a group of supporters invaded a meeting at the National Academy of Sciences devoted to rDNA science and policy. His strategy at the NAS is best described by the 1960s term "guerrilla theater," a dramatically staged political confrontation. Fearing major disruption of its conference, academy officials allotted Rifkin time at the start of the programme for a statement.

Meanwhile, his associates, donning silk stocking face masks, stood like motionless icons, projecting a grotesque apparition of human mutants. While Rifkin took over the podium of the academy auditorium, his colleagues stretched a banner across the stage with a quote from Adolf Hitler that stated: "We will create the perfect race."

On another occasion, Rifkin organized a signature campaign of scientists and religious leaders who were asked to support a statement against genetic manipulation of the human germline. The list of supporters included individuals of vastly different ideological persuasion. To have such a politically diverse group on the same side of an issue was itself a news event. It proved worthy of a front-page story in the *New York Times*. In the fall of 1983, when the RAC was reviewing a proposal for field-testing ice minus, Rifkin brought national attention to the problem of deliberate release by filing suit against NIH on the grounds that its approval was given prior to a full study of the environmental consequences.

His suit was supported by two organizations, Environmental Action and the Environmental Task Force. To strengthen his case, Rifkin received sworn affidavits from leading ecologists who opposed the release of the ice minus strains, at least until more could be learned about the environmental consequences. Rifkin's lawsuits have been rallying points for critical debate on genetics issues.

Philosophically, he personifies the modern genetics Luddite. But on a practical level, he has made effective use of environmental law to impose a greater burden of responsibility for assessing the risks and eliciting the ethical consequences of new technologies.

Ironically, he is secretly applauded by many who, while they might differ with him philosophically, find a certain degree of relief in slowing the pace of regulatory approval. Environmental groups, skeptical about deliberate release, have been hesitant to act more aggressively because the hazards of genetic engineering are still hypothetical. For Rifkin, it was less important to prove the hazards exist than to rally support against the hubris of science and reexamine its assumptions about progress.

There is a striking comparison between the old and new genetics controversies on the role of scientific participation. In 1976, the molecular geneticists were taken to task for not calling upon infectious disease epidemiologists to evaluate the risks of transforming *E. coli* into an epidemic pathogen. Subsequently, the infectious disease community did participate in the risk evaluation. Similarly, ecologists played no role in the rDNA debate until the controversy over deliberate release erupted.

Their entry into the discussion over deliberate release came when the social and political context was ripe for their participation. Ecologists are viewed as natural allies to environmentalists; when the latter became involved in the debate over field tests, they sought advice from the former. Also, the Committee on Science and Technology of the House of Representatives made specific recommendations to NIH and USDA that they revise the memberships of their advisory committees to include individuals specifically trained in ecology and environmental sciences.

Finally, the EPA began developing its own policy on deliberate release in the early 1980s. The agency is accustomed to working with ecologists, for example, in evaluating pest control methods. The first genetically engineered organisms planned for environmental release were classified as biological control agents. The EPA began hiring consultants to study the potential environmental problems of releasing such agents. A terrestrial ecologist from Oak Ridge National Laboratory, Frances Sharples, completed a study for the EPA in 1982 on the effects of the introduction of organisms with novel genotypes into the environment.

Microbial ecologist Martin Alexander of Cornell University chaired an EPA study group on biotechnology. In 1984, ecologists participated in an EPA sponsored workshop at the Banbury Centre at Cold Spring Harbor, New York, on the evolutionary consequences of biotechnology. Also, in 1985, with support from the EPA, the USDA, national environmental organizations, and corporations, ecologists, geneticists, public interest representatives, and federal regulators met for four days at Cornell's biological field station in Bridgeport, New York, to develop principles of containment for genetically engineered organisms tested under field conditions.

By raising the issues of biotechnology in the national media, Rifkin also helped catalyze the concerns and perspectives of ecologists on the problems of deliberate release, particularly those who had not been brought in as consultants to the major agencies. Eventually, letters from ecologists appeared in *Science,* followed by responses from geneticists. A dialogue was opened between molecular geneticists, microbiologists, and ecologists on their approaches to risk assessment. Soon it became evident that the issues over ice minus unfolded into a larger debate about scientific culture, epistemology, and disciplinary hegemony.

We will see how the ice minus bacterium became an international symbol for resistance against biotechnology by some groups like Earth First and the European Greens. To the biotechnology industry, ice minus had quite a different symbolic value as a product that could raise public confidence in a new technological frontier that promised to advance agricultural efficiency while making peace with nature.

5

The Environmental Treaty

MAKING SYSTEM

The complex interactions and procedures by which global environmental agreements are formulated, ratified, and implemented are rarely thought of as a system. But these agreements are, in fact, governed by predictable sets of actors engaged in a relatively structured process of negotiation, constrained by formal and informal rules and customs. The "system," such as it is, exists by default in many cases, and its weaknesses may not be apparent. Before the treaty-making process can be strengthened, we must understand the way this system operates.

The actors in the environmental treaty-making system include governmental leaders, unofficial or nongovernmental interest groups (including environmental action organizations, business associations, and scientific associations), and multilateral entities—particularly agencies of the United Nations such as the United Nations Environment Programme, the World Bank, and the United Nations Development Programme. These actors gather information, exchange ideas, formulate proposals, and meet in informal and formal sessions to negotiate, prepare legal documents, and vote whether or not to accept new responsibilities, including taxing themselves to cover the costs of monitoring their global environmental management efforts. They meet periodically to review how well they have done and determine whether or not to take further action.

For the most part, these interactions are guided by the formal rules and informal practices that the United Nations has evolved over several decades. They are also shaped, to a lesser extent, by a body of international law—mostly what is called "soft law"—that reflects commonly accepted norms. Finally, and most important, the treaty-making process is constrained by the global interplay of domestic politics: In the final analysis, only agreements that are politically acceptable to national leaders will be approved.

This complex web of actors, institutions, and practices is always in motion, animated by the efforts of individuals and organizations to advance their interests, selfish or otherwise. Not surprisingly, with all that is going on, it can be difficult to gauge, at any moment, whether or not the system is working well. Indeed, there are serious disagreements about how to measure the success of environmental treaty-making efforts. For those involved directly in treaty negotiations, the signing of formal agreements, after years of debate, seems like a tremendous victory. Yet, for the rest of us, it would be a mistake to measure success in terms of anything less than tangible environmental improvements, regardless of the amount of time or effort it took to hammer out the legal accords.

Many overlapping forces are at work in the biosphere, which makes it difficult to keep track of improvements (and what caused them). Indeed, because of the complexity of natural systems, scientists have great difficulty sorting out which actions account for which outcomes. We are only just beginning to understand global ecological interactions well enough to know exactly how seriously to take some of the threats that currently loom large. So, attributing specific changes to the requirements of particular treaties is almost impossible.

If we cannot attribute improvements to the implementation of specific agreements, how, then, shall we measure the success of global treaty making efforts? In large part, success is in the eye of the beholder. Those who have strong views about the causes of environmental problems, the remedies that will and will not work, and how responsibility for implementing agreements ought to be assigned see the results of treaty-making efforts in ways that reflect their biases.

Three debates, and where the participants stand on each, account for the most important differences in the ways that the success of international environmental negotiations is gauged. The first is between *pragmatists* and *idealists*. In this debate, both sides are concerned about environmental quality and sustainability, but they have conflicting expectations about what constitutes reasonable progress in the search for fair and efficient solutions to environmental problems. The second debate pits *optimists* versus *pessimists* in what seems like a never-ending battle over how to achieve global cooperation. The optimists and the pessimists are diametrically opposed on both the prospects and the range of global agreements that are possible.

The third debate is between *reformers* and *conservatives*. These two groups disagree about the desirability of restructuring the United Nations and the system of multilateral institutions that has evolved since the mid- 1940s. Where a person stands on each of these debates frequently determines the extent to which he or she thinks the UN sponsored, convention-protocol approach to environmental treaty-making is succeeding or failing.

Measure Success

For some environmental advocates, especially in less economically developed countries, merely getting official representatives together to talk about environmental issues is a sign of great progress. In these countries, little or no attention has been paid to local environmental problems, let alone to global environmental threats. For some environmentalists in these settings, the ones I call pragmatists, almost any effort—no matter how modest—is an important step in the right direction. Agreements can always be strengthened, they argue; the important thing is to get started.

On the other side are the idealists. They worry about treaties that "sound" good but yield few tangible improvements in environmental quality. In their view, these may be worse than no agreement at all. Empty promises, they assert, let politicians off the hook, allowing them to take credit for solving problems when, in fact, the environment may actually be deteriorating at a rapid rate.

Indeed, inadequate or partial agreements may forestall the efforts needed to achieve measurable improvement. Thus, the idealists maintain a stringent standard for assessing progress: measurable, documented improvements. All global efforts to deal with environmental problems ultimately hinge on the, willingness of agencies, organizations, and individuals to follow certain rules and, often, to change their behaviour.

Thus, it is not uncommon to evaluate environmental treaties in terms of the obligations that the signatory states promise to impose upon their industries, citizens, and governments. The pragmatists argue that even purely symbolic statements by a few countries are valuable because they put pressure on reluctant leaders who may be hesitant to make the most modest commitments. The idealists, however, are not satisfied with anything less than full fledged, enforceable promises to regulate environmentally destructive behaviour.

If progress on the environment ultimately depends on the willingness of countries to force citizens and enterprises to live up to tougher standards, the pragmatists argue, then even modest agreements provide support for insurgent environmental protection efforts. It is usually grassroots groups, after all, that shape public perceptions and impose pressure on government leaders to change their policies.

On the other hand, if the idealists are right, then the publicity accorded symbolic statements under scoring the need for action (but not requiring any) may actually undermine environmental protection efforts by leading potential supporters within a country to believe that problems have been solved, when indeed they have not. There are several reasons that signed international agreements (that appeal to the pragmatists) often produce little if any real improvement.

First, it often takes so long to secure international cooperation that environmental protection strategies that made sense when they were first proposed represent "too little, too late" by the time they are implemented. The problem may have reached entirely new (and very different) proportions in just a few years. For example, efforts to protect a particular habitat may be irrelevant once a species is extinct.

If too few countries ratify an agreement, the cumulative efforts of those living up to their promises may be insufficient to reverse the problem. It may be impossible, for instance, to clean up the ocean, even if most countries stop dumping toxic wastes, as long as a few countries refuse to halt their dangerous disposal practices.

It often costs more to implement environmental treaties than the signatories anticipated. Although they are listed among the signatories, some countries may actually renege when changing domestic priorities make it impossible for them to live up to their original intentions. Moreover, even after elaborate treaty language has been worked out, disagreements arise over what was intended and what was guaranteed. In the face of such disagreements, disgruntled countries sometimes decide to opt out. Occasionally, disagreements of this sort are nothing more than a cover for a country's change of heart when it discovers the true costs involved or encounters unexpected domestic opposition.

Sometimes, to get countries with significantly different needs or priorities to sign a treaty, a lowest-common-denominator or compromise approach is adopted. Such halfway agreements, not surprisingly, are often insufficient to achieve the intended results. Many times, the objectives of such treaties are laudable, but the programmatic commitments that countries were willing to make could not possibly achieve their espoused objectives.

Thus, even though agreements are signed, the results may be discouraging. Too few countries (or not enough of the right countries) may be involved. In other instances, even though a sufficient number of countries may be supportive at the outset, they sometimes fail to live up to their promises. In still other situations, even when most of the signatories are ready and willing to comply, the problem turns out to be worse than anyone thought, or the problem turns out not to yield to the solutions that were selected. For these reasons, the idealists refuse to count the number of treaties signed or the number of countries signing them as indicators of progress.

The 1991 UNEP Register of International Treaties and Other Agreements in the Field of the Environment and the 1992 Report of the UNCED Secretariat (entitled *The Effectiveness of International Environmental Agreements*, edited by Peter H. Sand) list more than 120 international agreements and international legal instruments in the environmental field. More than half are regional or sub-regional in scope (and thus are only partially relevant to the study of global agreements).

Of the approximately fifty global agreements listed in these two compendiums, about ten are statements of general environmental concerns, eight are concerned with civil liability for environmental damage (from oil spills or nuclear disasters), and another seven are International Labour Organization sponsored agreements on the protection of workers in the work environment. To be fair, several have slowed the rate of pollution or begun the process of protecting important natural resources, even if they have not repaired past damage. And the last few were signed only in June 1992 (and have not yet come into force).

Consider the dialogue between the idealists and the pragmatists that would take place over the treaties listed in Appendix A. The idealists would argue that most species of whales remain on the verge of extinction (although one or two have made a comeback). Moreover, several whaling nations have recently renewed commercial harvesting of whales after having previously agreed to phase it out. The rate at which wetlands are disappearing is increasing, and most damaged wetlands have not been repaired or replaced.

The idealists would probably point out that Antarctica is still not fully protected, and the countries that claim a share of the Antarctic have been unable to agree—even after twenty years of debate and study—on a permanent minerals mining ban. Further, many, if not most, of the significant ecological resources representing "the heritage of mankind" have not been protected from the adverse effects of development. Ocean dumping of toxic and hazardous waste continues while the sludge already at the bottom of the world's oceans has not been retrieved or treated.

From the idealists' standpoint, many endangered species and habitats continue to be lost each year, and there have been few successful efforts to replace those that have been destroyed. The level of hazardous and toxic waste produced each year continues to rise, and increasing volumes are transported from developed to developing nations, creating grave dangers for unsuspecting residents.

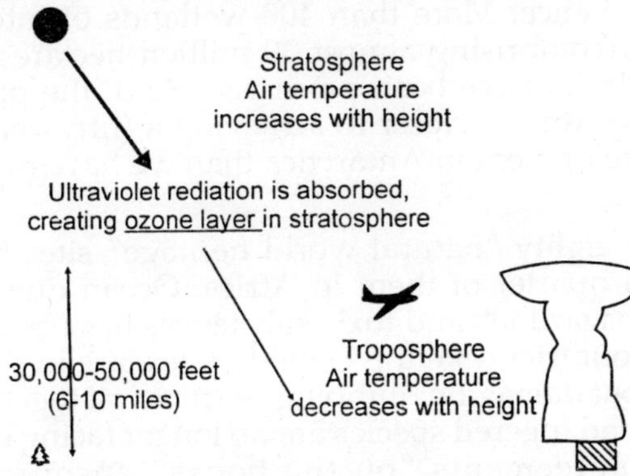

Fig. Ozone layer

The comprehensive Law of the Sea, negotiated for almost ten years, has not yet been (and may never be) ratified by enough countries to take effect. Some of the worst oil spills in history have occurred in recent years with little or no indication of any international cooperation to combat the harmful effects that these accidents have caused. And, although many countries have signed an agreement (called the Montreal Protocol) aimed at protecting the "ozone layer" by banning the use of chlorofluorocarbons (CFCs), a number of key CFC-producing countries have not signed; moreover, other equally damaging emissions, like methane, have not been restricted at all.

Finally, the climate change and biodiversity agreements signed at the Earth Summit in 1992 offer no guarantee that carbon dioxide levels will ever be reduced. In the absence of a treaty on forest protection, important rain forests will continue to be lost. Neither of the treaties signed at the Rio meeting includes specific standards that must be met or deadlines by which signatory countries can be held accountable.

The pragmatists have a very different story to tell. Countries that once paid no attention to natural resource management issues have made explicit commitments to do so. As a result, the environmental movement within each of these countries has been strengthened. The number of whales of all kinds is on the increase, and several have even reached the point where commercial whaling is once again viable.

International norms now suggest that it is important to recognize the vital role that wetlands play in maintaining ecological balance. More than 400 wetlands of international importance (comprising almost 30 million hectares—an area the size of Italy) have been preserved. And, the pragmatists would argue, we are closer to achieving a (fifty year) ban on mineral development in Antarctica than we have ever been in the past.

Almost eighty "natural world heritage" sites have been protected, a quarter of them in Africa. Ocean dumping of a number of hazardous and toxic substances has been cut back, and some countries, like the United States, are in the process of phasing out damaging dumping practices all together. Some previously endangered species are no longer facing extinction. There are agreements "on the books" encouraging the reporting and cleanup of oil spills. Migratory flyways are more clearly delineated and some have been protected by countries that previously ignored them.

The pragmatists point out that although the formal Law of the Sea treaty has not yet come into force, many of its provisions have, de facto, been incorporated into international law. Not only have efforts to protect the ozone layer led to a voluntary ban on CFC production in a number of countries but more and more countries are joining the phaseout, as financial arrangements to assist developing nations are finalized.

In addition, there is a system in place to keep better track of hazardous wastes transported across national borders.

Progress may be slower than some would like, but it is measurable. The fact is, we have climate change and biodiversity agreements that can be strengthened as additional scientific information becomes available.

Global Cooperation

Whether one sides with the pragmatists or the idealists, there are several reasons to be pessimistic about the prospects for achieving the level of cooperation required to manage shared (or common) resources like the oceans, space, Antarctica, the atmosphere, or the diversity of species. The first is the worsening split between the developed nations of the North and the developing nations of the South.

The second is the stubborn persistence of national sovereignty as an important goal unto itself. The third is an apparent lack of incentives sufficient to bring some nations to the bargaining table for serious discussions about the nature of global environmental threats and the challenges of sustainable development. The optimists believe that all three of these obstacles can be overcome, making international cooperation possible. The pessimists are doubtful.

North-South Conflict

The North-South split is often portrayed as a battle over money and technology, but there is more to this conflict than economic and scientific ascendancy. Some observers paint the nations of the South as a supplicant begging for additional aid while the North is portrayed as a wealthy but selfish benefactor unwilling to share its technological secrets. Ever since the 1972 Stockholm Conference on the Human Environment, when the developing nations managed to generate a conference-approved set of principles that challenged prevailing approaches to economic development and environmental protection, the North-South debate has intensified.

Often the disagreement revolves around whether funds for the implementation of environmental protection agreements will be added to the development assistance that is already provided to the South ("additionality"), and what (if any) strings the North will attach to these funds ("conditionality"). At the Earth Summit, much attention focused on the issue of technology transfer or technology sharing. U.S. biotechnology firms were worried that the Biodiversity Convention would require them to turn over products they might invent, using materials gathered in the South (even after paying an initial royalty). The South argued for continuing royalties and technology sharing.

The injustice of cultural hegemony (that is, the overwhelming impact of Western culture and the forces of modernization on economically dependent countries) undergirds the development assistance and technology transfer debates. The South wants the North to acknowledge the unfairness of this indirect form of domination. These debates mask the real source of conflict, which is a fundamental difference in how the nations of the North and the South think about progress.

This dichotomy is elegantly presented by Thijs de la Court in *Beyond Brundtland* (1990), in which he describes the Third World's response to the report of the UN World Commission on Environment and Development (otherwise known as the Brundtland Report, after the prime minister of Norway who chaired the commission). The disagreement, as he explains it, is really over the meaning and direction of economic development. Although it may be difficult, particularly for Americans, to believe that most of the developing world—if it had a choice—would prefer not to emulate contemporary Western development patterns, that is indeed the case.

To paraphrase two noted Indian critics of the Brundtland Report, there are other ways of defining progress: we need not equate development with economic growth, economic growth with expansion of the market economy, modernity with consumerism, and nonmarket economics with backwardness. The Brundtland Report (which popularized the idea of sustainable development and postulated the need to link economic development and environmental protection) assumes that effective responses to global environmental threats can be found within the framework of the current pattern of economic development, if only the key actors would accept the importance of sustainability. This is, in fact, the generally held view in the North.

The South, however, views its current array of problems (i.e., population growth, food shortages, the loss of forests, the difficulties of producing energy, the impacts of industrialization, and the burdens of massive urbanization) as by-products of the dominant economic development pattern. The South wants the North to accept responsibility for *causing* these problems by pursuing a form of economic growth and an approach to development that is fundamentally at odds with sustainability.

Even though the leaders of many developing nations are currently championing more market-oriented approaches to development and economic growth, they certainly are not seeking to obliterate the unique social, economic, and ecological conditions that make up their cultural identity. In the long run, such a course of action would be (among other things) self-defeating. So, if every nation sought to achieve the levels of per capita energy use and resource consumption currently enjoyed by the United States, the world's reserves would quickly be exhausted. There must be some other pattern of development, therefore, or some fairer way of sharing the world's resources that we ought to pursue instead.

Attitudes toward the World Bank (and other multilateral lending institutions) reflect these contradictory views about what constitutes desirable development. Although the World Bank announced in 1987 its intention to show greater concern for the ecological consequences of its investments than it had in the past (e.g., promising to prepare environmental impact assessments before making future funds available), it did not immediately curtail its support for many large-scale development projects that some in the South see as totally unacceptable.

Indeed, the bank's declaration that it would pay more attention to environmental quality was met with harsh skepticism by many Third World environmental action groups that have bitterly opposed bank-financed projects that they feel betray a fundamental blindness to anything other than the North's definition of desirable development. In Brazil, the

World Bank and the Inter-American Development Bank have supported the Palonoroeste Project aimed at opening up virgin tropical forests in the Amazon for new development.

In China, the World Bank-financed Three Gorges Dam on the Yangtze River will flood thousands of square miles and force three million people to relocate. In Botswana, the bank has supported large-scale cattle ranching projects that have caused overgrazing of ecologically vulnerable land. The bank's dam and irrigation projects in the Narmada Valley of India have been opposed bitterly by grass-roots groups trying to save their communities. And in Indonesia, the bank has supported the emigration of millions of people to unspoiled areas of other islands in the archipelago.

These projects, sometimes referred to as "the fatal five" by critics of the bank, symbolize for many in the developing world the North's continued unwillingness to honour the South's commitment to alternative patterns of development. This, in turn, feeds suspicions that the North's recent statements of concern about environmental quality and sustainability are nothing more than a pretext for further exploitation of the South.

The issues of technology sharing and development aid are far from unimportant, but they are secondary to this larger question. The issue is not how much more money the North will provide to the South, but whether the underlying North-South relationship can be shifted from one of dependence and confrontation to one of fruitful interdependence. In 1986, according to the United Nations and the Organization for Economic Cooperation and Development, the nations of Africa received $18 billion in development assistance.

In that same year, the same countries faced $34 billion in losses: $15 billion in loan redemption obligations and $19 billion in export price drops. At the time, Africa's total debt approached $200 billion, a sum that equaled half the continent's overall GNP and three to four times its annual income from exports. As the Brundtland Report explains: "Debts that they cannot pay force African nations relying on commodity sales to overuse their fragile soils, thus turning good land into desert. Trade barriers in the wealthy nations—and in many developing ones—make it hard for Africans to sell their goods for reasonable returns, putting yet more pressure on ecological systems." What does it matter, then, how much more aid or what kind of technology is made available if the level of dependence only increases and the impact on fragile ecosystems only worsens?

The North-South conflict will not be resolved by doling out additional money or making new technologies available on favorable terms. The South expects the North to accept a greater share of responsibility for the difficulties that developing nations face. The South is also waiting for the North to acknowledge that there must be a change in Northern lifestyles if greater fairness in the allocation of the world's resources is to be achieved. From the North's standpoint, neither demand is reasonable. Thus, the deadlock continues, especially because the nations of the South, also called the Group of Seventy-seven (even though there are more than 125 nations in the group), have found their voice and mobilized more effectively in recent years.

Sovereignty

In March of 1989, at The Hague, the prime ministers of France, Holland, and Norway proposed an ambitious plan to create a global environmental legislative body with the power to impose new environmental regulations and binding legal sanctions on any country that failed to carry them out. The proposal failed, although twenty-four heads of state did adopt a declaration calling for a new United Nations authority empowered to act even without unanimous agreement.

From time to time, perhaps more out of frustration than anything else, proposals like this one calling for the creation of supranational bodies with the power to override national sovereignty are put forward. They do not succeed because countries fight desperately to maintain their individual rights and privileges. Most global environmental agreements worked out through ad hoc negotiations include only weak monitoring and enforcement provisions.

This, too, is a function of national efforts to maintain not only control over all decisions within their geopolitical borders but autonomy over actions that affect common areas and resources as well. For example, the International Convention for the Regulation of Whaling established the International Whaling Commission to oversee the provisions of that treaty but failed to give it enforcement powers. So, treaty violations are noted and announced, but sanctions cannot be brought against nations that are in violation.

Indeed, if a country is upset about being charged with violations, it can threaten to "opt out" of the agreement or organize a rump group to set competing standards. Monitoring and enforcement powers are not granted because they appear to conflict with the prerogatives of national sovereignty. Without effective monitoring and enforcement, though, implementation of treaties is difficult. Most countries comply with most existing international agreements, but there are many instances of blatant disregard for rules and deadlines. Sovereignty is often used as an excuse. Countries that find themselves out of compliance assert that the more important issue is that their sovereignty is being undercut by other nations.

One school of international relations holds that because sovereign nations will always act in their own self-interest, international institutions are irrelevant. That is, such institutions will never be able to convince nations to pursue a course of action inconsistent with their self interest. A second school of thought believes that we don't need multilateral institutions because self-interested nations in a competitive setting will always work to achieve mutually beneficial exchanges without any prodding from an international body.

As Arthur Stein writes in Why Nations Cooperate, "Like a well-functioning market in which self-interested behaviour leads to optimal, efficient outcomes, an anarchic international system composed of self-interested states should need no regulation." However, many nations engaged in fairly regular efforts to formulate and implement cooperative arrangements create and empower new institutions to make their collaboration easier. They do this in spite of their desire not to give up their sovereignty.

The international relations theorists notwithstanding, sovereign states seeking to pursue their self-interest often realise that their ability to build and maintain cooperative relationships depends on their capacity to sustain appropriate institutional oversight and assistance. In the same way that communities of people form governments (and in the process give up a measure of autonomy in exchange for security), the nations of the world must, when they come together to work out ways of handling global problems, surrender some degree of sovereignty. The important point, though, is that they do this by choice. And, they can do it in one policy arena while deciding not to do so in another.

The 1982 Law of the Sea Treaty advanced the idea of a "common heritage of mankind" that would have diminished the zone of absolute sovereignty of individual states (i.e., whenever their actions threatened global environmental quality). This was a challenge to the prevailing legal assumption that sovereign states can do whatever they please outside the jurisdictions of other states. Because the Law of the Sea Treaty was not ratified, however, this limit on sovereignty was never adopted.

Of course, many nations long ago accepted other practical limitations on their sovereignty in order to partake of the advantages of international communications and trade. The operation of ports, airlines, telecommunications, and other global systems are all governed by international authorities. As countries find themselves increasingly economically interdependent, their sovereignty diminishes.

As new satellite technologies allow global monitoring without direct access to territory, traditional notions of sovereignty are further softened. Nevertheless, while notions of national sovereignty over the management of natural resources (both within a country's borders and in common areas) are evolving in response to technological and economic change, they still pose a substantial obstacle to effective environmental treaty making.

Incentives to Bargain

Finally, the pessimists are worried (and rightly so) that many nations will refuse to join in global environmental negotiations not because they are worried about losing their sovereignty but because they fail to see what they have to gain. These countries are potential "free riders," (i.e., parties that benefit by the actions of others without sharing any of the responsibility or cost). They assume that others will make enough of an effort so that they will benefit from an environmentally safer world without shouldering any of the costs. Why join the club if you can have all the benefits without any of the costs?

Many of the costs of combating various kinds of pollution must be paid now, although the full range of benefits generated by such pollution control efforts will probably not be realized until well into the next century. This presents a problem for many politicians. Their time horizon runs only until the next election. As with efforts to raise taxes, for example, few elected officials want to be the ones to initiate increases. They all, though, want to share the credit and the benefits when it turns out that the money was well spent. Many environmental treaty negotiations focus only on the allocation of costs (including constraints on development). They devote little or no attention on the tangible benefits that will be generated or how they will be shared.

The impacts of environmental problems and the costs of combating them will almost certainly not be distributed equally. Some countries will lose more than they gain—even in the long run. Unlike other international negotiations, in which the losers are guaranteed benefits of other kinds, loser countries have good reason to remain on the sidelines or even to sabotage international environmental treaty-making efforts.

Although linking treaty-making efforts together could change this calculus, it is rarely done. Environmental negotiations have been conducted largely in isolation from negotiations on other international issues such as debt, trade, or security. The complexity of linkage, according to the pessimists, would be beyond the capacity of international institutions to handle. On balance, these three obstacles—the North-South conflict, the desire to preserve national sovereignty, and the lack of incentives to come to the negotiation table—suggest that the pessimists have good reason to doubt that the future of environmental treaty making is bright. The optimists believe that those deficiencies can be corrected by adjusting the international legal system.

An Inadequate Legal Structure

The Brundtland Commission recommended the adoption of a universal declaration on environmental protection and sustainable development analogous to the Universal Declaration of Human Rights. This was aimed at overcoming one of the most serious inadequacies in the international legal system: the lack of specific national obligations to protect the environment. According to many commentators, particularly Phillipe Sands (director of the Centre of International Environmental Law in London), there are actually two key problems.

First, nongovernmental organizations and other "nonstate actors" have no standing in the international legal system (i.e., they are not recognized as legal persons). Only sovereign states are recognized, and they are all equal. It is not likely that one state will be able to prosecute other states on behalf of the global environment, and it is not permitted for nongovernmental actors to assume this prosecutorial role (i.e., negotiate with states and appear before international tribunals). Second, environmental rights have not been established on an international scale. That is what the Brundtland Report was attempting to do.

In the absence of a universal declaration on environmental protection and sustainable development, the Vienna

Convention on the Law of Treaties, governing the making of global environmental agreements, provides only process guidelines. The convention spells out some ground rules, although it leaves a great many questions unanswered.

Most multilateral environmental treaty negotiations are initiated by international organizations. In recent years, the UN Environment Programme (UNEP), has been the primary initiator, calling the conferences that produced the Convention on the Ozone Layer in 1985 and the Basel Convention on the Control of Transboundary Movements of Hazardous Wastes and Their Disposal in 1989. The Earth Summit, while officially under the aegis of a separate UN Conference on Environment and Development, was also a product, at least in part, of UNEP's efforts.

Other treaty-making efforts have been stimulated by small groups of countries or international scientific organizations like the International Union for the Conservation of Nature and Natural Resources (IUCN). The Vienna Convention on the Law of Treaties does not specify who should initiate treatymaking efforts. Some conferences involve only a few countries meeting on a subregional basis. Others may involve most of the 180-plus members of the United Nations.

Once a conference has been called and the negotiating committees for each nation have convened, they formulate rules regarding how their negotiations will proceed; specifically, how long the conference will last, who will be allowed to participate, who will present scientific evidence, how proposals will be made, and how the participants will formulate the text of an agreement. These rules may be suggested by the international organization that called the conference, or they may have been spelled out in a previous agreement.

The Vienna Convention does not specify which countries should participate or how the treaty-making process should proceed. It does say , though, that adoption of the text of a treaty requires "a vote of two-thirds of the states present and voting, unless by the same majority they decide to apply a different rule." Procedural rules can have a major impact on the chances of reaching agreement.

For example, the use of a "single-text" approach, in which there is only one draft of a potential agreement and all parties write their suggested changes on that draft rather than offer alternative versions of the whole text, has been cited by negotiation experts as an important factor in the success of several treaty negotiations. Similarly, the participation of nongovernmental organizations (which often provide technical information to formal delegations) aided in the successful negotiation of the Vienna Convention on the Protection of the Ozone Layer. By contrast, the limited duration of the Basel conference (on the transport of hazardous substances) has been cited as a reason that its treaty was unacceptable to some of the nations involved.

The Vienna Convention on the Law of Treaties neither suggests nor rules out a single text procedure. It offers no suggested timetable for negotiations. It also fails to address the issue of roles for nongovernmental organizations because, as already mentioned, such groups have no standing in international law. Once the parties have adopted the text of an agreement, the next step is to secure signatures. For sub-regional conferences involving a relatively small number of parties, each country normally signs at the conclusion of the conference.

For larger regional and global conferences, agreements usually "remain open for signature" for an extended period of time at one or more locations. For example, the Vienna Convention on the Protection of the Ozone Layer was open for signature in Vienna for six months and then in New York for six months. Sometimes agreements remain open for signature indefinitely.

This was the case with the Convention on Wetlands of International Importance Especially as Waterfowl Habitat (Ramsar). Signature is not a trivial step because parties unhappy with the adopted text of an agreement may refuse to sign it. This is especially true when a text is adopted by majority vote rather than by consensus. Once a party signs an agreement it must—according to the Vienna Convention on the Law of Treaties—refrain from activities that would defeat the objectives of the agreement. There are no sanctions prescribed for countries that violate this requirement.

The Vienna Convention provides that an agreement take effect, or enter into force, when a sufficient number of parties have agreed to be bound by it. For most multilateral agreements, the parties can choose to have the agreement enter into force when a fraction of the states have indicated that their governments have ratified it. For regional agreements involving a relatively small number of parties, agreements usually enter into force when all parties have ratified it. Large regional and global agreements typically set a minimum number of required ratifications.

The Climate Change Convention, for example, will enter into force when fifty of the 153 signatory countries ratify it. The Biodiversity Convention requires only thirty of the 152 signatories to ratify. The Ramsar agreement is an exception; it entered into force when seven nations became parties but allowed other nations to become signatories later. Once the minimum number of ratifications is achieved, the agreement enters into force only for those parties who ratify. There is no requirement, according to the Vienna Convention on the Law of Treaties, that a minimum number of countries participate in a treaty-making effort.

Multilateral agreements can be modified after they enter into force. Unless a treaty stipulates otherwise, the general rule is that all signatories must be notified of any proposed modifications. The parties then have the right to participate in modification negotiations, and to sign any subsequent agreement.

Although most multilateral agreements follow this pattern, there are exceptions. Some modifications may be binding on all parties to the original agreement. In such cases, a supermajority vote is usually required for the modification to be adopted. For example, the Vienna Ozone Convention allows any party to propose new amendments. It requires the parties to make "every effort" to reach agreement by consensus, but if they fail, the amendment may be adopted "as a last resort" by a three-fourths majority vote of the parties at the conference.

A typical treaty, or convention agreement, has a predictable set of headings or sections. Most begin with articles defining the key terms used in the agreement and specifies its geographic scope. Next, there are articles calling on the parties to take "all appropriate measures" to address the problem, cooperate with one another in promoting scientific research, share information, deal with emergencies, and carry out the other provisions of the agreement. Sometimes the agreement expressly calls for the parties to formulate specific accords, or protocols, on each of these items. In such cases, the convention outlines the provisions for establishing protocols.

Additional articles usually call for periodic meetings or follow-up conferences. At these meetings, delegates review new scientific information and establish additional research objectives; they assess the effectiveness of the individual and joint measures taken to combat the problem; and propose, discuss, and vote on additional protocols and amendments. Such meetings typically occur once every few years. "Extraordinary" meetings may be called at the request of a predetermined number of parties (sometimes one-third, sometimes half the signatories).

Other articles establish a secretariat (although the Vienna Convention on the Law of Treaties does not specify who should play this role). The primary duty of the secretariat is to call and supervise meetings. Additional responsibilities of the secretariat include transmitting information submitted by one party to the others, ensuring coordination with other international organizations, performing any functions assigned to it in the protocols, and preparing reports about its activities. The secretariat is usually an international organization such as UNEP.

Proposed protocols or amendments to a treaty must be submitted to the secretariat well in advance of regular meetings. The secretariat then sends them to the other parties several months before the next scheduled meeting. At these sessions, the parties try to reach agreement by consensus, but if they fail, amendments can be adopted by a vote of those present. These include technical definitions and actions appropriate to various special circumstances. For example, the annexes to the 1973 Convention on International Trade in Endangered Species of Wild Flora and Fauna (Cites) indicate which species are accorded various levels of protection. The annexes to the London Dumping Convention categorize various substances that may or may not be disposed of in the ocean.

Finally, most treaties contain articles defining how they will enter into force. Other articles determine when and how long the agreement will remain open for signature. Another will designate a depository whose duties are to receive notices of ratification and rejection and, when enough parties have ratified, to inform the governments involved. The depository is usually the government of the nation that hosted the meeting at which the convention was signed. Further articles can specify the number of countries that must ratify for an agreement to enter into force, and how long the agreement is expected to remain in force (usually indefinitely).

The reformers look at this legal structure and see glaring weaknesses: the rules are very sketchy; no one is really in charge; much of the negotiation process is ad hoc and unregulated; there is no central authority to manage the process or compel compliance; and the dispute resolution mechanisms available through the International Court of Justice are not definitive.

Conservatives, as I call them, view the treaty-making apparatus and nod approvingly: there is no overbearing bureaucracy telling countries what they must do; the fact that procedures are sketched only in broadbrush strokes means that it is relatively easy to make whatever adjustments are needed; and most countries seem to be in compliance most of the time. The contrast between these two views can be traced to the very differ ent levels of confidence that reformers and conservatives have in the mechanisms for making and enforcing international law.

According to the noted legal scholar Patricia Birnie, international law can be developed either by treaty, by custom, or by reference to general principles. Treaties are clear; customs are not. Custom, in international law, refers to practices that provide concrete evidence of a country's willingness to abide by certain rules. General principles of law are contained in the decisions of national and international tribunals. In addition, scholarly writings shape their interpretation. UN General Assembly resolutions and the resolutions of other UN bodies and conferences also shape world opinion about general principles of law.

The accommodation of interests by diplomacy has given rise to a legal order, even though there is no centralized legislative body that formulates or enforces international law. The reformers want a more explicit supranational system with the power to legislate and coerce uncooperative states; the conservatives not only hold out no such hope but oppose all efforts to eliminate sovereignty as we know it.

Because states are sovereign, their consent, or at least their acquiescence, is required to develop international law. Birnie suggests that acquiescence may be assumed if states do not protest at critical junctures, when new customs are crystallized, but this does not speak to the issue of compliance or enforcement. The International Court of Justice provides a dispute resolution mechanism that can sometimes be called into play, but it has no power to coerce a country that refuses to accept its jurisdiction or judgments. In sum, the international legal structure provides very few and very modest guidelines to regulate global environmental treaty making.

Fundamental Flaws

Most recent international environmental negotiations have followed a two-31 step approach. An initial series of meetings is held to review scientific evidence and draft a framework convention. Then, subsequent meetings of the signatories focus on the preparation of detailed protocols. The convention-protocol approach has produced a number of agreements, the Climate Change Convention and the Biodiversity Convention being the most recent examples of first-step agreements. The Montreal Protocol is the best-known example of what the second step in the two-step process is supposed to generate.

The Vienna Convention on Substances that Deplete the Ozone Layer was signed by twenty-one countries in 1985, eight years after the initial meetings were held to review scientific evidence concerning ozone depletion. The Montreal Protocol was signed two years later by twenty-seven countries, and took effect in 1989. The protocol was revised in 1990 at the first four-year review session in London. The signatories decided to amend the targets and timetables they had set earlier, and took up the question (again) of how to provide sufficient assistance to the nations of the South to enable them to meet the terms of the protocol.

The convention-protocol approach allows countries to "sign on" at the outset even if there is no agreement on the specific actions that must be taken. The 1975 Barcelona Convention (a regional agreement that produced the Mediterranean Action Plan) established procedures for monitoring various sources of pollution without ordering specific pollution controls or reduction levels.

Most countries could agree that further documentation of pollution levels would be desirable, but commitments to specific targets would have been difficult for some countries to accept because of domestic opposition to the short-term economic consequences. Nevertheless, the signing of this and other conventions created momentum and encouraged a commitment to continued scientific inquiry.

The 1992 Climate Change Convention sets no targets or timetables (even though many countries, particularly the Europeans, wanted them). Instead, subsequent meetings will consider the results of ongoing scientific inquiries and explore the possibility of adopting specific implementation measures. In the meantime, some nations have unilaterally adopted timetables and targets that affect only themselves.

Although these are not legally binding (that is, subsequent legislatures in those countries are free to abrogate them), they provide a benchmark for subsequent negotiations and may give the countries that have taken action the upper hand in negotiations over follow-up protocols in that they can declare that the standards they have already adopted should become the norm. Groups within countries concerned about environmental protection have been able to point to the signing of framework conventions as proof that further action is still required.

In some cases, the accumulation of scientific evidence (made possible by the signing of a convention) will be sufficient to melt political resistance to remedial actions. In other instances, the force of world opinion, stimulated in part by the signing of a convention, has been sufficient to pressure reluctant countries to sign agreements. Sometimes, with the mere passage of time, domestic opposition has weakened, making it easier for national leaders to build support for the actions outlined in follow-up protocols.

A major shortcoming of the convention-protocol approach, however, is that it encourages a process that is often long and drawn out. The 1973 Cites Convention was not signed until ten years after the IUCN first called attention to the need for an international effort to regulate the export, transit, and importation of endangered animal and plant species countries, and their products. During that decade, many animal and plant species were destroyed. Likewise, the Law of the Sea negotiations took ten years, during which time opportunities to preserve and develop maritime resources were lost.

Often the dynamics of the convention-protocol approach reinforce the tendency to seek lowest-common-denominator agreements. The Basel Convention on the Control of Transboundary Movements of Hazardous Wastes and Their Disposal, for instance, incorporated vague language and avoided the politically difficult task of defining key terms. This made it possible for reluctant countries to sign but undermined the chances of successful implementation.

Indeed, the developing nations of Africa met in 1991 to sign a regional accord called the Bamaco Agreement, going far beyond the terms of the Basel Convention, to ban the importation of hazardous waste by African countries. The Basel Convention calls for the disposal of hazardous waste in an "environmentally sound manner," but it does not say what this means. The standard of environmental soundness is left entirely up to each country. The Bamaco Agreement was an effort on the part of some African nations to postulate tougher environmental standards.

In a 1989 meeting of the signatories to the Cites Convention, the group agreed to recategorize the African elephant, shifting it to a list requiring more stringent protection. The proposal was resisted by certain ivory-trading nations and was accepted only after a provision was added allowing individual countries to petition for a "down listing" of the African elephant from endangered to threatened. The criteria for determining the appropriateness of down listing were not specified in the regulations. In the final analysis, this might well thwart the effectiveness of the treaty.

Most international environmental treaties impose the same requirements on all signatories. Indeed, that is why the lowest-common denominator solution is often the only viable option. The Basel Convention regulates the movement of hazardous wastes between signatory countries but allows for bilateral agreements between signatory countries and nonsignatory countries. This explicitly contradicts an earlier provision of the agreement that states that signatory countries cannot ship hazardous wastes to nonsignatory countries. The provision allowing bilateral agreements can certainly be interpreted as a watering down of the overall treaty; however, its inclusion was politically necessary to hold the agreement together.

Obviously, all treaty negotiations require a give-and-take among countries. In the case of environmental agreements, however, merely satisfying the political demands of the countries involved is not enough. The dynamics of the natural systems involved must be respected, regardless of the political implications. Another key problem is that agreements produced by the convention protocol approach sometimes reflect an outright neglect of available scientific and technical information. They sometimes incorporate requirements that turn out to be technically infeasible or illogical.

In negotiating a pollution control strategy for the Mediterranean, for example, political considerations overshadowed the technical wisdom of including the Black Sea states of Bulgaria and Romania. According to Peter Haas, author of *Saving the Mediterranean*, a regional agreement was made to satisfy political pressures even though all the parties involved knew they were doing the wrong thing from the standpoint of common sense.

Unfortunately, when it comes time to work out specific second-step protocols, the terms of the original convention can get in the way of producing technically appropriate agreements. Some observers have noted that U.S. support for regulating CFCs was first constrained, and later facilitated, by the development of CFC substitutes by Dupont Chemical Corporation, the major CFC producer in the United States.

During convention negotiations, countries will often cite scientific evidence that justifies the general policies they prefer. Because the implications of these policies need not be confronted at that point, counterproductive policies, which frequently constrain subsequent protocol design, may be adopted. This is a very real danger of the two-step drafting process.

Finally, negotiations that occur through the convention-protocol process are often dominated by the most powerful states. The final deal leading to the Montreal Protocol was negotiated primarily between the United States and the European Economic Community, the two largest consumers and producers of CFCs. The developing world was kept on the sidelines. African states have called the Basel Convention a "sellout" by Third World signatories because of the inability of the weaker states to win inclusion of provisions shifting the liability for hazardous waste disposal from recipient states to the generating or exporting states.

The ad hoc nature of the convention-protocol approach to treaty making is partly to blame. In the absence of a formal negotiation system spelling out the rights of each country to help set agendas and the obligations of the secretariat to achieve a certain minimum threshold of support from UN members before environmental treaties can be considered, the most powerful states can, in effect, write the rules, control the dissemination of technical information, and dominate the negotiation process.

The ad hoc convention-protocol approach as currently practiced also fails to come to grips with important negotiation problems. For example, the convention-protocol approach actually encourages the "hard bargaining" tendencies of many countries in the sense that it does little to discourage countries from misrepresenting their interests. It is not de signed to encourage countries to separate the tasks of "creating options for mutual gain" from the task of securing agreement. Furthermore, it focuses insufficient attention on building informal agreements and coalitions prior to formal meetings. These are well-known negotiation problems that contribute to the difficulties of all multilateral negotiations.

Countries often misrepresent or exaggerate their needs as part of their bargaining strategy. This tactic may be useful in one-shot, single-issue negotiations; however, in circumstances where long-term working relationships are crucial to implementation, the parties involved are better off if they make only statements that turn out to be true. This leads to greater trust, which makes implementation easier. In addition, when negotiators are in the business of trading concessions (rather than engaging in a search for trades that maximize joint gains), they must keep checking back with their leaders at home. Very little creativity is possible under these circumstances.

An inordinate amount of time is wasted performing the "concessionary dance." In sum, when negotiations involve many issues and many parties who will have to deal with one another on a continuing basis, it makes more sense to share candid accounts of each side's interests and avoid positional bargaining. There is nothing in the two-step approach to environmental treaty making that pushes the parties in this direction.

When negotiations are conducted without a clear separation between creating multiple options for dealing with specific issues and choosing among them, negotiators usually fail to explore the full range of possibilities. A willingness to explore options, or what mediators often call "supposals," is too often misconstrued as a commitment because the parties fail to distinguish "inventing" from "committing," as Roger Fisher, William Ury, and Bruce Patton suggest in *Getting to Yes*.

This inhibits creative problem solving. All too often, negotiations become a test of will. When they do, parties lock into a battle over a small number of options, which closes down creative brainstorming (that might yield additional alternatives responsive to the interests of all). The two-step structure of the convention-protocol approach makes no distinction between the tasks of inventing and committing, although it could. At the convention-writing stage, the goal is typically to stay at a very general level so that all countries will at least agree that some (unspecified) action is needed to solve a problem.

At the protocol-writing stage, the goal is usually to find a formula that everyone can accept. Such formulas typically contain a great many exceptions, underscoring the fact that no country can be forced to sign. Convention and protocol writing are usually treated as zero-sum games; that is, everyone acts as if anything good for one country must be bad for another. By the time countries have come together to negotiate treaty language, they have usually locked into positions. When agreements do emerge, they are usually the result of compromise by the most powerful parties rather than the result of creative resolution of differences.

In many of the examples cited above, the agenda for each negotiation was set through informal communication between the secretariat or lead organization and a few dominant nations. For example, in the meetings of the International Whaling Commission, the agenda was rigidly controlled by the whaling nations for many years. Certainly, in negotiations in which each nation knows its self-interest, the agenda should be crafted to ensure that the issues of greatest concern to affected states are included.

If important issues are omitted, some nations will have no incentive to participate, or they may feel compelled to sabotage the negotiations or subsequent implementation of agreements. However, a more flexible and inclusive approach to agenda setting for treaty negotiation is needed. Unfortunately, the narrowing of negotiations that occurs when specific protocols are debated one at a time, and the general formula is already set by a signed framework convention push environmental treaty negotiations in exactly the wrong direction.

The convention-protocol approach to international environmental negotiations as currently practiced also fails to take account of the special qualities of environmental problems. Many negotiations tend to focus solely on allocating losses or costs incurred through environmental regulation but do not deal with the gains resulting from wiser resource management, or the ways in which these gains might be shared.

Indeed, most environmental treaties aim to curtail pollution or regulate the use of common resources by restricting the activities of participating countries. The economic losses suggested by these new rules provide solid excuses for many nations not to join the negotiations. Ways of sharing the economic and ecological benefits of environmental protection are rarely offered as a compelling rationale for participation, although the gains really ought to be the focus, not the losses.

In addition, environmental issues almost invariably involve a degree of scientific uncertainty that complicates decision making. Our understanding of the natural world is incomplete. Forecasting tools provide only crude approximations based on often unverifiable assumptions. Disagreement among technical experts on complex environmental issues is pervasive. If one country thinks it will be disadvantaged by a particular policy proposal, it can easily locate sympathetic experts to raise doubts about the adequacy of the scientific evidence put forward by others.

If a country wants to delay implementation of costly pollution abatement measures, it is not hard to argue that further study would be desirable before long-term commitments are made. At the same time, the dynamics of environmental deterioration are hard to reverse. Even if there were an immediate 100 percent cut in ozone depleters worldwide, for instance, it would still take approximately sixty years for chlorine concentrations in the atmosphere to come down to their 1985 levels. It is also worth pointing out, as certain environmental advocacy groups have, that 1985 levels were the levels that permitted the hole in the ozone layer to occur!

Many environmental issues also hinge on the problem of how best to manage a common resource, or its converse, how to penalize "free riders"—the beneficiaries of environmental improvements who refuse to pay for or help achieves them. If the benefits of regulated development or pollution control are diffuse, and accrue to all nations irrespective of their behaviour, some nations will feel little or no incentive to accept any restrictions. For example, nations such as India have refused to sign the Montreal Protocol. Reductions in the use and production of CFCs by other nations (especially bigger users) will slow down the rate of ozone depletion, and India, as well as the rest of the world, will benefit.

The tragedy, however, is that if most countries took this stand, it would never be possible to curb ozone depletion. There is nothing about the convention protocol approach as currently practiced that solves the free-rider problem or motivates collective action. Environmental negotiations, up to now, have been conducted largely in isolation from negotiations on other international issues such as debt, trade, or security. Negotiations sponsored solely by UNEP cannot speak to linkages between environmentally related actions and other important economic and security-related considerations.

Recently, some developing nations have raised the desirability of making these linkages, particularly between environmental protection and international trade. Environmental scientists, certainly, wish that linkages among different environmental problem-solving efforts could be made more explicit. What ecologists call a "cross-media approach" is necessary to ensure that the solutions to one problem do not merely shift the risk or the impact to another domain (where the effects might even be worse).

Finally, the convention-protocol approach as currently practiced fails to respond to the need for effective monitoring and enforcement. Most environmental agreements worked out through ad hoc international negotiations include only weak monitoring and enforcement provisions. Almost every agreement listed in Appendix A relies on self-reporting and self-enforcement.

Monitoring and enforcement are difficult because they conflict, as was pointed out earlier, with the prerogatives of national sovereignty. Yet, without effective monitoring and enforcement, implementation of any agreement is difficult. Although it is true that there is a relatively high rate of compliance with most of the terms of most international treaties, it does not take much noncompliance to undermine the effectiveness of a treaty. And, there has certainly been noncompliance with many global environmental treaties. All of these weaknesses were clearly in evidence in the Earth Summit negotiations.

The Earth Summit

The 1972 Stockholm Conference on the Human Environment marked an important turning point. Before 1972, separate UN agencies dealt with environmental problems on a piecemeal basis. After Stockholm, the United Nations moved to coordinate its environmental activities through the creation of the United Nations Environment Programme, or UNEP. UNEP is not an executive agency. It is not financed or empowered to carry out its own programmes.

Instead, it depends on other national and international agencies to implement the programmes it helps to design. Along with the Brundtland Commission and a variety of other international organizations, UNEP encouraged the United Nations General Assembly (in 1989) to create the United Nations Conference on the Environment and Development, or UNCED. The UNCED charter declared: "The Conference should elaborate strategies and measure to halt and reverse the effects of environmental degradation in the context of strengthened national and international efforts to promote sustainable and environmentally sound development in all countries."

As late as 1991, the major results of the Earth Summit were expected to be four signed conventions on climate change, biological diversity, biotechnology, and forests (down from seven as originally mandated); a charter of planetary rights defining basic guidelines for balancing environmental protection and economic development (otherwise known as an "Earth Charter"); a plan of action for setting the planet on a path toward sustainable development in the twenty-first century (Agenda 21); a redefinition of roles and responsibilities for various UN agencies; agreement on the financial mechanisms needed to implement Agenda 21; and a general agreement on technology transfer.

Three working groups were established by the UNCED Secretariat, headed by a Canadian, Maurice Strong (who also headed the Stockholm Conference). The first working group was chaired by a Swede, Ambassador Bo Kjellen. It focused on protection of the atmosphere (i.e., climate change, depletion of the ozone layer, and transboundary air pollution); protection and management of land resources (i.e., deforestation, desertification, drought, and land degradation); conservation of biological diversity; and environmentally sound management of biotechnology.

The second working group was headed by Dr. Bukar Shaib of Nigeria. It, too, had an ambitious agenda: protection of oceans, seas, and coastal areas, and the planned use of their living resources; protection of the supply and quality of fresh water resources; environmentally sound management of hazardous wastes and the prevention of illegal international traffic in toxic waste and other dangerous products; environmentally sound management of toxic chemicals; and improvement of the living and working environment of the poor, protection of human health conditions, and improvement in the quality of life. The third working group was chaired by Bedrich Moldan from the former Czechoslovakia.

This group (which was not established until almost a year after the first two) was responsible for the legal, financial, and institutional framework for an "Earth Charter" and Agenda 21. In addition, the third working group had responsibility for such cross-cutting questions as the financing of international programmes; removing barriers to technology transfer; design of new legal institutions and instruments; possible reforms in the organization of the UN agency structure; economic instruments like pricing policies, tradable permits, national economic accounting systems, and fiscal incentives; penalties to encourage enforcement; and supporting measures like information management, training, and public education.

Four preparatory meetings (PrepComs) were held. The PrepCom membership was drawn from the 150-plus national delegations committed to participating in the Earth Summit. Many delegates were drawn from missions based in Geneva (because of the UNCED Secretariat's location there) and New York, site of the General Assembly. Delegations also included national technical experts. Nongovernmental organizations participated both as part of national delegations and as official observers.

The first PrepCom in Nairobi, Kenya, in August 1990, resulted in a request to the UNCED Secretariat to prepare eighty (later reduced to thirty) background reports. There was a great deal of wrangling over the role that nongovernmental organizations should be allowed to play in UNCED's deliberations. At the second PrepCom in Geneva, in March 1991, another fifty reports were requested. This avalanche of requests was an indication that the PrepCom participants were unprepared to deal with the difficult issues that had been put before them. The principal achievement of PrepCom II was the creation of Working Group III.

PrepCom III was held in August 1991, in Geneva, and was much more successful. There was agreement that there should be an Agenda 21. Most of the countries present supported the idea of an "Earth Charter." A preliminary statement of general principles on the protection of the world's forests was also approved.

PrepCom IV was held just before the Earth Summit, in March 1992, in New York. Although the session snagged on the question of financial resources, those present adopted and sent along to the Rio conference a draft of Agenda 21. On the final day of PrepCom IV, the Group of Seventy-seven broke off negotiations on a financing plan, leaving the specifics to the negotiators in Rio.

The Group of Seventy-seven was working from a draft on finances prepared by China and Pakistan. It called for "adequate, new and additional funds, covering the full incremental costs with no reallocation of existing multilateral or bilateral financial flows" to finance Agenda 21 activities by developing countries. The Group of Seventy-seven also wanted a special fund for the implementation of Agenda 21 that would (1) be in addition to official development assistance already targeted to developed countries; (2) require mandatory contributions from developed countries; (3) give equal voice to all parties in determining project eligibility criteria, project selection, and the release of funds; and (4) fund activities according to the priorities and needs of the developing countries.

Not surprisingly, the industrialized countries bracketed all of these demands and offered alternatives on each point.

The PrepCom chair, Tommy Koh of Singapore, directed a small group to try to work something out but, after two weeks of behind-closed-door meetings, there was no progress. Koh then tried to formulate a compromise, but it, too, fell apart when the European Community insisted that only the Global Environmental Facility (GEF)—a temporary creation of the World Bank, UNEP, and the United Nations Development Programme (UNDP)—should allocate the funds to implement Agenda 21.

The Principles on World Forests were drafted by the PrepCom, but the Climate Change Conventions and Biodiversity Convention were prepared independently by Intergovernmental Negotiating Committees established several years earlier by UNEP. Subcommittees involving representatives from more than one hundred countries struggled to find treaty language that would be acceptable at Rio.

When the actual Earth Summit was finally held in Rio de Janeiro in early June 1992, more heads of state were in attendance than at any previous UN conference on any subject, although not for the full two weeks. U.S. President Bush held off deciding whether to attend until just a month before the conference, unconvinced that the interests of the United States would be well served but worried about the impact of nonattendance on his standing in the presidential election. Numerous resolutions were introduced in the U.S. House and the Senate, not only calling on the president to attend but spelling out—in no uncertain terms—U.S. support for the strongest possible versions of all the treaties scheduled for discussion.

In the end, Bush attended, but for just three days. The United States was roundly criticized in the world press when the president refused to sign the Biodiversity Convention. The United States was also criticized when, at the last minute, Bush tried to push his own version of a forest protection treaty. He failed to commit substantial new and additional development assistance, and, in general, played no visible leadership role.

The conference was not able to agree on an "Earth Charter" establishing new legal rights or responsibilities relative to environmental protection or sustainable development. The final version of the Rio Statement calls on all countries to do everything possible to promote sustainable development but breaks no new ground on this subject.

Although the United States refused to sign the Biodiversity Convention, 153 other countries did. The convention calls for all countries to prepare policies, programmes, and plans to conserve and protect biological diversity. Each signatory is supposed to do everything it can to ensure access to its genetic resources as well as to technologies ("for sustainable use of biological diversity") that will help the other signatories. The developing nations are promised the full increment of aid required to meet the terms of the treaty (although amounts are not specified and mechanisms for distribution must be worked out at a later date).

The United States agreed to sign the Climate Change Convention, but only after it was watered down substantially. Just months before the Earth Summit, the European Community was adamant that it would not support a climate change convention that did not include mandated targets and timetables for reductions in the emission of greenhouse gases. (Many European nations had already adopted substantial carbon dioxide reductions on their own.) It is not clear how the United States maneuvered the Europeans into changing their stand, especially because many Group of Seventy-seven members were quite willing to isolate the United States by proceeding with a treaty that the United States could not sign.

The Climate Change Convention calls on all countries to do all they can to mitigate climate change by reducing or preventing the emission of greenhouse gases, promoting sustainable development, cooperating in preparing for adaptation to the impacts of climate change, and exchanging relevant scientific information. The North is called upon to "adopt national policies and take corresponding measures on the mitigation of climate change, by limiting its anthropogenic emissions of greenhouse gases and protecting and enhancing its greenhouse gas sinks and reservoirs."

Again, the developed countries promised to provide the South with "new and additional financial resources to meet the agreed full costs incurred in complying with the treaty." Presumably, as additional scientific evidence becomes available to support the claim that global warming is indeed occurring, and with the reports in hand required by the Climate Change Convention, it will be possible to specify timetables and targets in subsequent protocols. A great deal of attention at Rio was focused on the level of aid that the North would provide to help implement Agenda 21 and the two conventions. (There was no agreement on a forest treaty or a treaty on desert expansion.) Maurice Strong announced before the start of the Earth Summit that it might take as much as $125 billion a year to implement Agenda 21.

It was not clear how much new money was actually promised at Rio, but even the most optimistic total falls far short of Strong's estimate. The developing world uses 0.7 percent of Gross National Product as an overseas development assistance target for the nations of the North. A few Northern countries accept this goal, although very few have actually met it; most, including the United States, still do not endorse this target.

The money that the North promised will flow, in large part, through the Global Environmental Facility (GEF), even though the Group of Seventy-seven was not entirely happy about this. Some accommodation was made to the South when the organizers of the GEF (the World Bank, the UNDP, and UNEP) agreed to change its administrative structure somewhat to ensure more of a role for the South in making allocation decisions. This was a response to the charge that the United States (through its supposedly disproportionate control over the World Bank) had too much say over GEF allocations in developing countries. The issue of control remains unsettled.

Agenda 21 grew longer and longer as the Earth Summit preparations proceeded. By the time everyone had finished adding to it, the document was almost eight hundred pages in length and included more than forty chapters plus appendices. It is not really an agenda for action because it contains no priorities of any kind.

The results of the Earth Summit and the events preceding it offer conclusive evidence of the weaknesses of the existing environmental treaty-making system. There was, after all, no international agreement on the range or content of the environmental threats that had to be addressed. UNEP, and then the UNCED Secretariat determined the process rules; others went along. There was little or no philosophical agreement on the types of solutions likely to be most effective, or even on the most efficient ways of allocating responsibility for implementation.

The idealists would say that the Earth Summit failed miserably because no ironclad commitments were made to reverse or repair environmental deterioration. Rio did not produce responses to many of the serious environmental threats we now face. The pragmatists would say, though, that substantial progress was made: two framework conventions were signed, and there were resolutions adopted on a great many other issues, denoting a new level of international attention to problems that must be recognized before collective action is possible. Additional funding was promised by the North, and a great deal of public attention was focused on the idea of sustainable development.

Although North-South differences did not block all action, they certainly bounded the debate and made it impossible for the United States to sign the Biodiversity Convention. Almost nothing was done to strengthen multilateral institutions with global environmental management responsibilities. There was agreement to create the United Nations Commission on Sustainable Development, but its powers will be extremely limited. No new law was written—nothing like an Earth Charter or the earlier Brundtland Sustainable Development Declaration was enacted. We will have to wait several more years to see whether follow-up protocols on climate change and biodiversity are enacted, and whether there will be full-fledged forest and desert treaties.

We certainly can consider the agreements reached at Rio as lowest common-denominator results. The countries involved politicized the search for scientific understanding, engaging in some of the worst adversary science ever seen. They minimized the search for creative options by focusing narrowly rather than broadly on the full range of environment and development issues requiring attention. They undervalued the importance of benefit sharing, focusing more on short-term economic costs than on long-term environmental gains for future generations.

There were no innovative proposals for ensuring compliance or dealing with free-rider nations. The final results were dominated by the most powerful states, while nongovernmental interests were left to formulate unofficial treaties at the parallel Global Forum for nongovernmental organizations. I am certain that the system that produced such modest results at UNCED can be strengthened, even in the face of worldwide recession, continued North-South hostility, and a fierce determination to preserve national sovereignty.

6

Biotechnology with Green Credentials

BIOTECHNOLOGY IMPLICATIONS

Modern biotechnologies have truly revolutionary implications. They confer a *generic* technical capacity to undertake selective genetic intervention in existing forms of life and to create novel life forms. The genetic code can now be manipulated and nature refashioned according to the logic of the market place. As a result, biotechnologies have suddenly new alternatives and paths of development for all major actors in the food system: farmers and input suppliers, primary processors, final food manufacturer and customers. It is this crosscutting, polyvalent capacity of biotechnologies which makes them such a potent force in the food system.

This kind of optimism about the potential and promise of genotechnology is repeated in the field of medicine where a parallel and similarly wide-ranging development is taking place. The genome project which seeks to map the entirety of the human genome, genetic diagnosis of hereditary diseases, and the potential for future gene therapy excite not just medical scientists but most of the major transnational bio-chemical companies, the financial markets, and the world of insurance.

Numerous collaborative ventures between academic scientists and business have evolved, making some of the participants millionaires before they have brought a single product on the market. Moreover, all the major chemical giants are involved. Thus, for example, Zeneca, the bioscience arm of ICI, employs 33,000 people worldwide and, according to Geoff Tansey and Tony Worsley, in 1992 had sales of nearly £4 billion. The scientific interaction with the basis of heredity is big business: the stakes are high, regulation meagre, public information virtually non-existent.

Genotechnology is the latest in a string of scientific developments that extend industrial societies' uncontrolled reach into the very, very long-term future. Like the chemical and nuclear technologies before it, genetic engineering is bringing about substantial changes in human existence which extends well beyond the time-space confines of industrial societies. It affects the future of glocal public health, agriculture and food production, trade and export, insurance and financial markets and, as patenting and the seed monopolies take root, the livelihood of peoples the world over.

It implicates city dwellers and contributors to the rural economy, members of industrial and tribal societies. It has an impact on peoples of both the Northern and Southern hemisphere. Like previous technological innovations, it holds out the promise of cornucopia: the end to food shortages and world hunger, poverty and disease, weather and season dependence, as well as the beginning of limitless supplies of food, new forms of speciation and genetic diversity, control over the processes of life.

As with the preceding technologies, the promoters of genotechnology focus on the promise of profit and control at the expense of issues associated with the potential long-term effects, on function and product at the expense of re/productive system processes-the visible *Merkwelt* without the invisible *Wirkwelt*. As before, therefore, the tip of the iceberg is mistakenly assumed to be the whole.

GENOTECHNOLOGY TO THE RESCUE

In the context of the discussions presented in this book, this newest and most lucrative branch of scientific research and its products impacts on glocal society, inter/national politics, economic activity, medicine, and agricultural practice in a number of significant ways: first, this technology relates to social stratification and the creation of differential life chances. Genotechnology, it needs to be appreciated, is redrawing the lines of inequality.

'Firms involved in biotechnology', as Geoff Tansey and Tony Worsley argue, 'promote a new form of enclosure, no longer of common land but of the common heritage of humankind in the genes of plants and animals which are needed in much genetic engineering work'. By patenting the methods and procedures of genotechnology, the owners of patents have effectively negated the farmers' ancient right to the means of reproduction. Saving seeds from one year to the next and thus ensuring the best growing potential for specific localities, for example, is no longer an option: the introduction of hybrid seeds meant that growers cannot do it; the patenting of genetically engineered seeds means that they are not allowed to.

This shift in practice marks a move from sustainable self-sufficiency to one of dependency on the products and services of agri-chemical and bio-tech companies. The transcendence of self-sufficiency, a second effect of genotechnology, is in fact one of the proud achievements of this technology: the mode and means of reproduction have been effectively appropriated by science and business.

Even though it has been a tradition in most countries thar a farmer can save seed from his [sic] own crop, it is under the changing circumstances not equitable that a farmer can use this seed and grow a commercial crop out of it without payment of royalty...eed industry will have to fight hard for a better kind of protection.

No company would be willing to part with what they took years and spent millions of dollars developing. It's a question of intellectual property rights. Each and every generation of plants and livestock together with appropriate chemical aids have to be purchased from their proud developers and owners-from Monsanto and Ely Lilly, Zeneca and Shell, Upjohn and Bayer, Hoechst and Ciba Geigy, Nestlé and DuPont-not to be reproduced but used, consumed and/or discarded.

Genetic engineering is therefore an issue of time-based control not so much over the mode and means of production but even more so over the mode and means of reproduction. The patenting of seeds, semen, plants and life processes, fertility treatment for women, and genetic testing in medicine all point in the same direction: the attempt to control the future through reproduction. Loss of self-sufficiency and power over the beings, people and processes involved happens to be one of the key by-products of this control.

A third impact of genetic design in medicine, agriculture and food production relates to the econo-political ideal of choice. Much effort is currently expended to force producers to declare all genetically engineered materials fully, especially those that enter the food chain. The issue of choice, however, is not exhausted by the indisputable need to have genetically modified products clearly labelled. Money and knowledge determine whether or not citizens are able to exercise choice. I have made the case with reference to money and the capacity to choose healthy foods.

Here I want to add the issue of knowledge. Since choice presupposes appropriate knowledge upon which to make decisions between alternatives, proper choice about labelled genetically engineered foods, genetic screening, and future genetic therapy can only be exercised after a substantial level of gene literacy has been achieved in any given population. In a context of widespread ignorance, in other words, choice is rendered meaningless. Loss of choice is thus inescapably tied to the genotechnology rhetoric of choice. There is a need, therefore, for people to become gene literate and for this knowledge to be as widely available and socially dispersed as computer literacy is today.

A fourth effect relates to the issue of scientific control. Genotechnology as the crowning glory of biotechnology developments brings to fruition the quest for technological rationalisation and control of all spheres of socio-environmental life. That is to say, with genotechnology science has extended its reach beyond the breadth of social life and natural phenomena to the depth of the life force of nature. It has transcended the product barrier and now operates at the level of nature's processes. This means scientific exploration is no longer tied to the *Merkwelt* of visible outcomes-the sphere of *natura naturata*-but has extended its investigative prowess to the *Wirkwelt* of nature's immanent processes-the realm of *natura naturans.*

That is to say, with gene manipulation, scientists are able to interact directly with the source of life and change. As such, genotechnology could be a means of closing the gap between technology and living processes. This, however, is not the case. Far from closing the gap, this paragon of Newtonian science, as I shall indicate below, turns the crevice into an unbridgeable chasm. The time-based critique offered throughout this book, therefore, becomes more pertinent than ever.

The image and rhetoric associated with this technology is clearly very different from the four effects I have just mentioned. Not the intensification of inequality, profit potential and control-over nature's processes, agricultural means of reproduction, and citizens' capacity to exercise choice-but the horn of plenty, increased diversity and biocentricity are the publicly hailed benefits of genetically engineered products. Cornucopia is the promise, especially where genotechnology is applied to the production of food, seeds and animal feeds. Science is to provide the technological means by which the potential is to be materialised, by which humanity's needs are to be furnished.

That is to say, science in the form of biotechnology and genotechnology tends to be promoted as the panacea to most of the major problems of twentieth century human existence: to hunger and food shortages, pollution and environmental degradation, the reduction in biodiversity, disease and even ageing. This technology, moreover, is presented as the scientific knowledge revolution that can repair the damage wrought by industrial activity associated with the use of machines, fossil fuels and synthetic chemicals. It has already produced micro-organisms that devour oil slicks; plants that thrive in communion with pesticides, herbicides and fungicides; foods that taste fresh after months of transportation and storage.

This technology, moreover, so the official line goes, is not merely the mode and means to solve societies' problems but it can deal with them in a safe and nature-friendly way. It is the soft, energy conscious path to a sustainable future. In the words of Earl H. Harbison, Jr., president of Monsanto which is one of America's foremost transnational companies involved in the development of biotechnology, 'we can at last find biological solutions to biological problems that mechanization and chemicals cannot solve'. Biocentrism, sustainable growth and a technology in harmony with nature are, however, difficult to achieve within the competitive system of global trade. Moreover, it does not square with the view of nature that arises from the scientific knowledge associated with this technology.

De-naturalisation and Economic Efficiency

Where transnational biotechnology and genotechnology companies wax lyrical about their green credentials, their biocentrism, and their environmentally sensitive way of exploiting nature, the scientists at the forefront of this development dream of control, mastery and the creation of nature to human design. Biocentrism as the *modus operandi* that is most in tune with nature's own ways is quite definitely not on molecular biologists' agenda. Nothing could be further from their intentions. On the contrary, a biocentric approach would bring their work to an immediate and abrupt halt: genetic engineers would have to reconsider their positions.

That is to say, the science that creates the basis for commercial exploitation of the deep-structure processes of nature is by definition intensely and exclusively anthropocentric: its aims are, after all, the *improvement* and the *redesign* of nature to human desire, requirement and convenience. Its vision is not one of harmony with nature but of an 'eighth day of creation', of taking over where god(s) and evolution have left off, of gaining knowledge and control, of improving on existing designs, re/creating nature for the exclusive needs of humans and their descendents.

This human-centredness applies whether the objects of research are animals that produce human proteins for pharmaceutical companies, genetically altered pig hearts for transplantation into human beings, pigs with an extra rib to provide more chops per carcass, a growth hormone that increases the milk yield per cow per temporal unit of measurement, a food crop that can be produced outside its natural habitat, or a recombinant virus that is to act as a vaccine against rabies in foxes. Every single development is exclusively for the purpose of human convenience, well-being, and/or profit enhancement. Questions about species being or the needs, rights and well-being of non-human life forms as well as nature and the environment more generally are definitely not under consideration here. They do not feature on the agenda of either molecular science or the biotechnology industry.

On the contrary, cornucopia, the future state of plenty, is achievable only by an explicit distancing from what is 'natural'. Through intervention at the molecular level, genotechnology accomplishes precisely that which is *not* possible in nature: the recombination of genes across breeds and species; the development within species and breeds of characteristics and capacities that do not and are not likely to occur naturally; the production of food outside its 'natural' context of existence, that is, independent from soil, water, photosynthesis, sexual reproduction, and the rhythmicity of seasons; the test-tube creation of body 'parts' which means their production outside living cells and bodies; the 'cure' of diseases before they happen. Sheldon Krimsky describes this denaturalising process with reference to the production of food:

> Food will be viewed as the output in a production process that changes a feedstock into proteins, vitamins, and carbohydrates, and creates a product that people or animals

either require for growth and nourishment or desire or would consume. There are many ways that these conditions can be met without farming the land.... The fermenter, bacteria, engineered genes and feedstock would replace soil, sun, rainfall, and germ plasm as the primary factors in production.

Not harmony with nature but distance, decontextualisation, disembodiment and the transcendence of nature's evolved processes of speciation and reproduction are the goal of this technology. Thus, instead of a science in tune with nature, for Jeremy Rifkin, 'genetic engineering represents the ultimate negation of nature'. Only in this form is it of interest to industry. Only from their 'unnatural' bases can genetically modified products develop into ultimate expressions of industrialisation.

Furthermore, not nature but the factory and industry are the metaphors and models that guide research associated with genetic manipulation. There is talk of 'gene machines' and genes as 'survival machines', of cells as 'chemical factories' and proteins as 'building blocks', of speed-up, efficiency and productivity. The machine metaphor indicates interchangeable parts-geno-technology rearranges bits of genetic information. *The cell as the material basis of life becomes the source for a new bacterial mode of industrial production.*

Industrial scientists study how to amplify or enhance the cell's products by introducing multiple copies of a gene by controlling environmental factors. As a 'factory', the cell must be brought into peak efficiency. In food production, such 'efficiency' is linked to an optimal conversion ratio of input to output such as grass and/or feed stuff to meat. In factory farming, this ratio has now reached its peak and is currently in a phase of diminishing returns where marginal increases in efficiency have reached a plateau. In the history of agriculture, each time such a phase had been reached, new technologies were invented to inject a boost of profitability: mechanisation, chemical assistance, economy of scale, and now genetic engineering. In this latest phase of intensive agriculture, the morphological features of animals and plants are redesigned so as to yield a more profitable conversion ratio.

Examples of such genetically engineered efficiency would be fowl consisting primarily of breast meat and crop plants with accelerated, extended or additional growth cycles for the more efficient use of land. In order to fully appreciate the significance of this technology, we need to grasp some of its defining features and the specific temporalities they entail.

Omnipotence and the Hazard Potential of Success

In everyday understanding, genes are associated with the part of heredity that is pregiven and predates individual existence; they thus stand for that which cannot be altered by upbringing. Molecular biology since the 1970s, however, has fundamentally altered, even negated this preconception. Far from being fixed, genes have become the sites for change of what was previously considered impossible even within the time frame of evolution: with the recombination of DNA sequences genotechnology has been able to transcend breed and species boundaries. Context has been rendered irrelevant. Quantity is no longer tied to sexual reproduction. Several changes can be achieved simultaneously rather than sequentially.

Genetic engineering as the deliberate manipulation of hereditary material is, of course, not a new but an ancient activity. As selective breeding, it has been practised almost as long as humans have been domesticating animals. What differentiates contemporary molecular genetics from this established tradition is its capacity to effect change at the level not of the phenotype but of individual genes. With the development of this radically new set of techniques scientists are able to intervene at the very source of being by decoding, comparing, excising, splicing, recombining, transferring, and cloning individual genes and sections of DNA.

This scientific manipulation at the level of genes opens up the foundation of life to human intervention and design. It allows for the injection of characteristics that are not part of the hereditary genetic make-up and enables scientists to combine morphological and functional characteristics that have evolved separately for millions of years and therefore do not occur anywhere in nature.

In their synthesis these human creations constitute truly new forms that can be replicated at will and in unlimited quantities. At that level, therefore, scientific dreams about omnipotence have become reality. At another level, however, the loss of control is unprecedented. The novelty of genetically engineered life-forms and their transcendence of nature's evolved processes and complex temporal interdependencies are also the basis for the magnitude of their inherent hazard potential. Thus, the increase in scientific control seems to stand in an inverse relation to the of control over effects.

The work of Regine Kollek, a German molecular biologist, is helpful for understanding this complex interdependency. The scientific manipulation of genetic material at the level of the genotype, she explains, is only possible because way back in the evolution of life-forms all organisms share generic origins and are thus genetically related. This means that even very distant species share nucleic acid sequences or functional genes so that specific proteins and enzymes, for example, may be found in the cells of yeasts, insects and mammals.

Only on the basis of this shared evolutionary prehistory, therefore, is the transfer of genes from one breed and species to another possible. What is crucial to note, however, is that proteins or enzymes with identical biochemical properties do not necessarily fulfil the same function, that is to say, the same material property has, depending on its specific context, a different ecological relevance. The invisible time-space of interaction is central to particular material expressions. The function of a specific gene, therefore, is not only given by its DNA sequence but by its 'location within a particular chromosomal and cellular context'.

Moreover, beyond this internal influence of cellular context and position, phenotypes are developing with reference to their extra-cellular contexts in interaction with their environments. This means that any genotechnological intervention has ramifications at all these levels. Parameters related to space, time, biology and natural history which influence the characteristics of individual organisms as well as the way they interact with other organisms, and which have proven themselves to be useful, perhaps even necessary, to life on earth, are therefore altered by genetic engineering techniques.

The controlled conditions of this experimental science create a context within which to produce specific, desired and highly predictable results about the action and function of *individual genes* only. Since, however, the multiple layers and levels of interactions and contextual differences do not form part of the experimental paradigm, the embodied, embedded phenotypic and environmental outcome is anything but predictable or certain. There exists no comprehensive theory to date, argues Regine Kollek, that is 'capable of describing the relationship between the functional effect of a gene and its spatial arrangement within the genome'. This means that the timescape of genes is only indirectly encoded in DNA and cannot therefore be deduced from the structure of individual genes.

Moreover, since interaction, interrelationship and socio-temporal context do not form part of the theoretical framework of the experimental science of molecular genetic engineering, its knowledge is only valid for the laboratory context. It does not apply for the interactive reality of phenotypic involvement in the environment. In other words, the behaviour of a genetically altered organism in its environment cannot be deduced or predicted from its controlled laboratory conditions.

Thus, whilst releases of genetically modified organisms, for example, may not necessarily result in a hazardous situation, the efficiency of the experimental manipulation nevertheless inescapably stands in an inverse relation to its predictability in the environment. 'Uncertainty and risk', Regine Kollek therefore suggests, are 'the price which must be paid for the total accessibility and control of these living objects in the laboratory'.

This genetic work at the 'leading edge' of science and technology with its associated creation of financial profits is thus inextricably tied to indeterminacy and, as I show below, the potential for hazard creation. Dreams of certainty and control are simultaneously within reach and utterly impossible. Mastery and omnipotence, in contrast, are unlikely for even the most tightly delimited laboratory context.

The everyday understanding which associates genes with the pregiven may not, after all, be fully negated. Focus on the temporal complexity of genotechnology gives us an indication why this might be so, why certainties and assurances need to be re-scrutinised, and why both theory and practice of the industrial way of life are in need of revision. The timescape perspective enables us to see the invisible. It facilitates recognition that the hazard potential of genetic engineering is inescapably and proportionally related to this technology's successes.

The accomplishments, we need to appreciate, are staggering. But, then, so are the perils: millions of years of co-evolution are circumvented. Reproduction cycles are dramatically speeded up or cut out altogether. Here the extraordinary achievement is counterbalanced by the inability to observe when things go wrong and to stop at each of the stages along the way. The achievement thus entails the loss of the established method of trial and error. Although the unprecedented speed-up of processes has obvious economic advantages, it clearly contradicts some very important scientific principles. Instead of exercising caution and precaution, genetic engineering is, in the imagery of Colborn *et al.*, like piloting a jet plane with a blindfold on.

Finely tuned, interdependent processes that can be abstracted from their context and set aside to be speeded up and slowed down at will allow for control over time and the temporal processes of nature. In a social system where time is money, such control is a tremendous feat. Its pursuit is irresistible, the potential for huge profits persuasive. Since, however, the ongoing system processes of interactive, mutual effectivity are unbounded, openended, long-term, time-distantiated, and often with long periods of immanence and latency, the hazard potential is matched only by the scale of the potential financial gains.

With the transcendence of the rhythmic and cyclical reproductive processes of life and death, growth and decay, production is freed to a substantial extent from its dependence on seasons and climate. Yet this achievement too is realised at a cost. It entails the transformation from an ecological, nowaste, recycling, reproduction system into an industrial linear, one-use, input-output production structure in which neither the production residues not the products after their use are reabsorbed as productive forces into the system.

They become 'waste'. As waste, the material not only contributes to the overall entropy of the system but it also interacts with its environment and moves invisibly through the networked system: unplanned, unsupervised, and uncontrolled. This current genetic control over life and death processes therefore is inextricably tied to the dramatic rise in hazard potential now and for the very long-term future, to the shaping of indeterminate, hazardous futures. This intensification of hazard potential needs to be acknowledged as real even though genotechnology has thus far escaped any recognised major and spectacular disasters.

'Promise' of Environmental Hazards

The genotech future, we are assured, is in the global society's best interest: genetic engineering is the only means to provide unlimited amounts of cheap food for ever-rising numbers of people. With nuclear technology the same argument was made about energy. Today the energy promise shows little hope of being fulfilled whereas the neglected hazard potential and the cost of decommissioning of this technology have risen beyond the innovators' worst nightmare scenarios. Genetic engineering, we are told, is the only means to satisfy not just today's but future socio-environmental needs in a safe and environmentally benign way. This too was promised by the promoters of nuclear power and the chemical management of the environment. It too turned out to be an unfulfilled and unfulfillable dream.

In addition to these similarities between genotechnology and predecessor technologies, however, there are a number of pertinent differences that require our attention. Whatever accidental changes may have been brought about so far by genetically engineered products, they lack the drama, immediacy and certainty of a Bhopal or Chernobyl accident. This absence of attributable and/or publicly acknowledged disasters tends to get confused with safety. Yet, as I have shown in this analysis, it does not mean that the hazards do not exist. It does not signify safety and should under no circumstances lull citizens into a false sense of security.

Like nuclear and chemical materials, genetically modified 'products' act invisibly at the level of cells. But their effects and symptoms, other than those intended by the specific genetic manipulation, are likely to be subtle rather than dramatic and/or not show themselves for many years. Moreover, the genetically altered materials' freedom vastly extends that of predecessor innovations. As living processes they multiply, mutate, migrate, and spread invisibly and unnoticeably by means that fall outside the remit of their creation. Organisms can turn malignant and develop into a pandemic without this being immediately obvious, noticeable or detectable. Quite clearly, this dramatically increases the materials' hazard potential.

With genetically modified organisms (GMO's), for example, harmful effects are not just more difficult to predict but even more difficult to trace to their origins than is the case with radiation, hormone disrupting chemicals, or diseases such as BSE and CJD. With thousands of genetically engineered viruses, bacteria, plants and animals developed over the last twenty years, it is now a case of potential hazards in the wings, of the global community of beings-present and future-having to wait and see what physical and socio-environmental 'effects' will ensue over the next years, decades, centuries, millennia.

When symptoms do arise there will be inescapable uncertainty about their causes. Proof will be unobtainable, a thing of the past, a luxury of a bygone era. The only certainty in this context of indeterminacy is the fact that released GMOs cannot be recalled, that they cannot even be traced. They have become free agents, instantaneous new life forms, ready to interact with others whose evolutionary pedigree extends over millennia. Opponents of genotechnology tend to view genetically modified 'products' as scientific monstrosities, as modern Frankensteins. Proponents, on the other hand, prefer to liken GMOs to the colonial period when plants and animals were brought back from foreign places and introduced into a new context.

While this Kew Gardens image of GMOs is intended to allay any fears and reservations people might have, it may equally conjure up memories of mutant killer bees and numerous other introductions of species from foreign lands that turned into disasters. The lessons to be learnt from such introductions of alien species are unambiguously clear: species co-evolve with other species and their environments as ecosystems. There are no cut-off points for co-evolutions. *The temporal web of interdependencies reaches to the beginning and the end of time.*

If you take something out of such a system, then all its other dependencies are implicated. If you add something, there is no way of predicting how that which has not co-evolved into a network of interdependencies will fare as an abstract, decontextualised, instantaneous entity that has to establish for itself a context for future networked relations of co-evolution. The organisms' past behaviour and response patterns may or may not be appropriate for its present and future survival. It may thrive, take over, go under or barely manage to exist.

Introduction of a new organism into the environment is precisely what genotechnology is about. Through manipulation at the level of genes this scientific skill and technique is used to develop novel micro-organisms for remedial purposes: for breaking down oil slicks and digesting crude oil; for combating rabies in foxes and other such designated threats to human health; for dealing with toxic chemicals in the environment and on waste dumps, even explosive substances and radioactive waste. While most of the hazards and their 'treatments' are real enough, there is no way of telling whether or not the 'remedy' will in due course become yet another item on the ever increasing list of hazards.

The Social Hazard

Hazards of a different, that is, a socio-political kind are created with the patenting of genetically modified organisms that range from microorganisms to plants and animals, including their parts and processes. The current drive towards patenting emanates from the US, with European and other industrialised countries not trailing too far behind. According to the European Patent Office, only genetically fixed plants are excluded from potential patenting. This means that it is just a question of time before all plants and animals are 'protected' under some patent or other, that is, before the entirety of natural reproduction has come under the control of a few transnational corporations.

To Vandana Shiva, 'this is just another way of stating that global monopoly over agriculture and food systems should be handed over as a right to multinational corporations'. It is the commodification and privatisation of life itself. As such it is not merely redrawing the lines of inequality but reviving some exceedingly unsavoury socio-political and medical practices, especially where these have the potential to be combined. An example would be the conjoining of patenting and cloning.

At present, humans are excluded from patenting and cloning. Yet, the processes if not the patents and policies are in place to screen for defects and disability and to clone humans with the most desirable and coveted traits. The scientific and medical skills are in place to ensure a future of perfect people, a future where all those who might turn out to be a drain on collective resources can be excluded either from insurance cover or, more sensible still, from existence. We seem to have been here before. In a previous political system, people with Huntington's disease, mental disabilities and a host of other 'abnormalities' ended up in the gas chamber or at the very least had enforced sterilisations.

Freely translated from Sabine Rosenbladt, the idea was that 'the collective had to be freed from bad hereditary traits, that the cost was too high for society to support an army of diseased, feeble-minded, cripples, deaf, blind and criminals'. For some forty years there was a moratorium on such ideas but now, it seems, the distance of one generation is sufficiently large not to associate the current brave new world of genotechnology-based 'preventative' medicine with the political system that developed such social eugenics to perfection.

With genetically engineered foods and medicines too, the practices and patterns of exploitation have a familiar ring. As with previous food experiments and introductions of new medical technologies, the powerless have been chosen to act as guinea pigs for the latest genetically engineered inventions: to test the 'cures', grow the plants, rear the animals, drink their milk, eat the food. With their substantially lower life-expectancy, the people testing these products could conveniently succumb to hunger, disease or strife long before the effects of those experimental foods and medicines show up as symptoms. This means, with a bit of luck-for the companies 'testing' their products-people and laboratory animals will be dead before the genetically modified nourishments and cures emerge from the latent stage as symptom, that is, before they are recognised as pathogens.

What, we need to ask, can testing mean in such a time-space distantiated context? How many generations, of what beings, over what spatial range, have to be clear of adverse symptoms for a product to be safe? How are latent and immanent multiplications, mutations, syntheses and migrations to be established and measured? How are pathogens to be traced to their origins? How is preclinical harm to be identified? And so the socio-scientific, socio-legal and socio-political questions go on.

Where genetically modified foods are concerned, the stakes are high in a dual way. First, genotechnology holds out the promise to grow artificially in the North what currently is provided by the South. Second, the scale of financial investments by transnational corporations and the stock market in genetic research and development is unprecedented. Disguised as food for the rising populations of the future, the business giants of the industrialised North are currently busying themselves with the creation of crops that will substitute cash crops grown at present by the South. They are developing plants that can survive harsher climatic conditions: designer plants that get genetically transformed into substitutes for existing crops such as alternatives to sugar; rape-seed oil that gets changed into olive, palm or coconut oil; palm oil that, in turn, gets redesigned into cocoa-butter. There seems in principle no limit to the possible mutations.

The impact of successful production of substitutes will be felt most by countries which have, in an earlier international division of labour, been made dependent upon export of the natural products concerned. This will be particularly destructive to the economies of Africa which depend entirely on single crops for most of their export earnings. While historically Africa was used to growing crops needed for Europe, in the emerging world order based on new biotechnologies, Africa will become dispensable as the North finds biotech substitutes for African crops.

This development should not be taken as a sudden conversion by transnational corporations: they are not redeveloping national loyalties. It is simply yet another innovative way to create new business opportunities and to pursue the competitive edge and financial fitness. Genotechnology, I want to suggest, is far more difficult to resist and fight than was/is nuclear power. Almost irrespective of whether or not there is wide-ranging public recognition of its dangers, genotechnology is being pursued to some indefinite bitter future (rather than end).

The reasons behind this are the size of the financial investment and the fact that, unlike the development of the nuclear sector, most of this investment and production is private or corporate. Although sensitive to consumer pressure, corporations are neither dependent on the vote for their survival not bound to national boundaries for their activities and outlets. Moreover, as decades of development come to 'fruition', any kind of retraction is out of the question, off the agenda, not open for negotiation. The result has to work at all cost as experience with earlier so-called wonder drugs such as human insulin, clotting factor and growth hormone has shown.

The failure of biotechnology was out of the question whether or not there were existing needs or favourable markets. The techniques would be used over and over again until successful products emerged. If the first generation of products proved unsuccessful or only marginally successful, other needs would be found to support a second generation of products.

The cancers and deaths that arose from some of these treatments were temporally close and direct enough for scientists and doctors to establish causal links. No such connections, however, are likely for the long-term changes at the molecular and genetic level as they work their way through bodies and systems. The problems that are recognised tend to be those of a previous technology. It seems a mark of true innovation that the real dangers associated with it still fall outside the involved scientists' imaginative capacity.

Moreover, the extremely mechanistic assumptions underpinning genotechnology predispose its activists against being able and/or inclined to conceive of the potential effects of this technology's application in post-Newtonian terms. Thus, I agree with Regine Kollek when she argues that 'the methods are both megaphone and hearing aid of scientific research'. This clearly is a difficult context for policy and regulation. Governments operating within frameworks of four to five year election cycles, concerned to boost their country's GNP and export, and keen to see jobs for the future are naturally drawn to the lure of this technology.

It promises to fulfil all those political needs and more. And it is safe-within the time-scale of politics, that is. Moreover, most major forms of regulation relating to health hazards and environmental issues arise from enforced responses to disasters or crises. The high level of invisibility and the absence of dramatic accidents or large-scale disasters, therefore, make genotechnology a poor candidate for regulation.

Initially, unfamiliarity brought about cautious restrictions. With the dramatic increase in genotechnological activity, its massive contribution to the national economy, and its assumed 'safety record', these early concerns are being gradually but surely whittled away. However, some regulating authorities are more inclined than others to apply the precautionary principle. The EU is an interesting case in point. One of the least democratically constituted regulatory structures, the EU, is actually more cautious than most other Liberal Democracies.

The continued moratorium on growth hormones in milk is a case in point. Here the precautionary principle is applied for fear of collapse of the dairy and meat market. This throws up an interesting paradox: whilst concerned citizens and environmental pressure groups are still conceived as agitators and trouble-makers, the threat of a consumer boycott seems to work wonders.

True to economic form and the ideology that underpins the globally constituted Liberal Democracies, who lead the way with their commitment to genotechnological developments, consumers are taken seriously by their political leaders. Globally constituted transnational corporations, in contrast, can ignore nationally located consumers as long as they are assured of the continued commitment by politicians, scientists, and the medical profession to the genotechnology cause. Together, of course, governments and transnational corporations have nothing but society's best interest at heart.

Finally, in the absence of major disasters, genotechnology has not had the breadth of press coverage and the depth of media analysis that would be necessary to facilitate the beginning of the kind of public genetic literacy I mentioned earlier. Information, analysis and debate at the popular level is virtually non-existent. Neither media hyperactivity around the cloning of sheep not academic debates late at night qualify for the sort of public dissemination of information and debate that is necessary for citizens to grasp the changes that are currently being set in place in the name of humanity's future.

At the moment, only the people who are directly affected are forced into an almost instantaneous 'coming up to speed' with the technology, its potential, limits and dangers. Offered genetic counselling or diagnosis of self, relatives or offspring, for example, recipients of this attention cannot avoid engagement with issues that will affect their long-term future. Faced with a decision of whether or not to get involved in hormone treatment of their children or livestock, people are forced to find out as much as they can before they come to a decision.

It is extremely unlikely that they would get any of that information from the news media, especially from the press, since only exotic claims, monstrous results, or disastrous effects will claim journalists' focused attention. Only in that form is the mediagenecy of genotechnology assured. Television and radio, in the UK at least, have offered some investigative programmes on food and medical issues surrounding the new technology but these were few and far between, late at night and on elite channels-certainly not enough to create widespread social awareness.

Moreover, science programmes such as 'Tomorrow's World' which tend to marvel uncritically at science's achievements are more likely than not to have strengthened expectations that genotechnologies are on the verge of providing the solutions to all current ills and troubles, and to all future needs of humans and 'their' environment. On the basis of the above, it may look as if hazard-creation was an inevitable, foregone conclusion, but this is not the case.

A change of direction may be extremely difficult to achieve, but it is certainly not impossible, and the temporal angle of vision gives us previously unnoted pointers to alternative paths. It certainly leaves us in no doubt that the problems and disasters associated with previous technological innovations cannot provide the necessary and sufficient knowledge about the kind of problems that will arise from these latest developments.

At the same time, their previous unpredictability and their continuous production of 'surprises' leaves no doubt about the need for caution and precaution for socio-environmental situations where the predictability afforded by the laboratory abstraction is lost and/or forfeited at the very moment the creation is set free to interact, transact, migrate, and move at an indeterminate, variable pace through the system as a whole. Thus, not the knowledge of the past, not the certainties of a previous age, not the science of proof and certainty, but proper engagement with temporality, indeterminacy and time-space distantiation have to become key issues for praxis.

In additon, leaps of the imagination are necessary for any meaningful projection into the future of systems that have outgrown the scientific framework of their creators. In this book I have not merely put a different spin on 'the environmental problem' but through the timescape perspective I have sought to make the invisible visible, the im/material real. Focus on the time-space of environmental hazards helped to reformulate both 'problems' and 'cures' and guided attention to the processes below the surface, to the *Wirkwelt* of *natura naturans* that acts immanently for indefinite periods and over unspecified places. It enabled us to redefine the nature and interdependence of safety, risks and hazards, to recognise and articulate the need for new conceptual tools, metaphors, theories and information structures. It made real what is currently assumed non-existent. As such, the timescape perspective has the potential to facilitate the will to a sense and sensibility that befits the current hazard re/production of the industrial way of life.

Genetically Modified Crops

Perhaps no human activity is more critical to human well-being and environmental quality than agriculture. Having sufficient food and adequate nutrition is the first step to better health; lower mortality rates; and a longer, more productive, and more fulfilling life. At the same time, agriculture has literally shaped the world's landscape, rivers, and waterways. And its effects are often felt in estuaries, in oceans, and even in the atmosphere far from where the actual agricultural activities take place.

Agriculture is the human activity that has had the greatest effect on the world's biological diversity. Today, agriculture alone accounts for 38 percent of global land area (FAO 2001), 66 percent of water withdrawals, and 85 percent of consumptive use worldwide. It is also the major determinant of land clearance and habitat loss worldwide. Between 1980 and 1995, developing countries lost 190 million hectares of forest cover mainly because their increase in agricultural productivity was exceeded by growth in the demand for food; in the same period developed countries increased their forest cover by 20 million hectares because their productivity outpaced demand (FAO 1997).

Agriculture and, to a lesser extent, forestry also affect biodiversity through water pollution and atmospheric transport—for example, by release of excess nutrients, pesticides, and silt into the environment. Demand for agricultural and forest products is almost certainly going to increase substantially. The world's population will almost inevitably grow from about 6 billion today to perhaps between 10 and 11 billion in 2100, an increase of 70 to 80 percent.

The average person is likely to be richer, which ought to increase food demand per capita. It is probable that the predominant future environmental and natural resource challenge facing the globe will be the problem of meeting the human demand for food, nutrition, fiber, timber, and other natural resource products while conserving habitat and biodiversity. The question is whether biotechnology can help or hinder the reconciliation of these often opposing goals. Although most of the following discussion focuses on agriculture, with particular emphasis on developing countries, much of it is equally valid for other human activities that use land and water, such as forestry, and for developed countries as well.

Environmental Benefits of Bioengineered Crops

According to 1998 UN projections, global population, which hit 6 billion in 1999, will most likely grow to 8.9 billion in 2050. This figure assumes that global crop production per capita will grow at the same rate between 1997 and 2050 as it did between 1961 and 1963 and 1996 and 1998, and that new cropland will, on average, be as productive as existing cropland in 1997 (an optimistic assumption).

If the average productivity in 2050 is the same as it was in 1997— hardly a foregone conclusion—the entire increase in production (106 percent under the above assumptions of growth in population and food demand) will have to come from an expansion in global cropland. That would translate into additional habitat loss of at least 1,600 million hectares beyond the 1,510 million hectares devoted to cropland in 1997 (FAO 2000). Much of that expansion would necessarily come at the expense of forested areas. It would lead to massive habitat loss and fragmentation, and put severe pressure on the world's remaining biodiversity and on *in situ* conservation.

On the other hand, a productivity increase of 1 percent per year, equivalent to a cumulative 69 percent increase from 1997 to 2050, would reduce the net amount of new cropland required to meet future demand to 325 million hectares. Such an increase in productivity is theoretically possible without resorting to biotechnology, provided sufficient investments are made in human capital and research and development, extension services, infrastructure expansion (to bring new land, where needed, into production and integrate it with the rest of the world's agricultural system), inputs such as fertilizers and pesticides, and the acquisition and operation of technologies to limit or mitigate environmental impacts of agriculture.

A 1 percent per year increase in the net productivity of the food and agricultural sector (per unit area) is within the bounds of historical experience: it increased 2 percent per year between 1961 and 1963 and 1996 and 1998. What is more important, there are many existing-but-underused opportunities to enhance productivity in an environmentally sound manner, but they are underused largely because of insufficient wealth—one reason why cereal yields are usually lower in poorer nations.

Merely increasing the 1996–98 average cereal yields in developing and transition nations to the level attained by Belgium-Luxembourg (the country group that had the highest average yield) would have increased global production by 141 percent; increasing the average global cereal yield to that typical in Belgium-Luxembourg would increase global cereal production by 163 percent. Notably, the theoretical maximum yield is 13.4 T/ha or 350 percent greater than the average global cereal yield in 1996–98.

Specifically, conventional (i.e., nonbioengineering) methods could be used to increase net productivity in the food and agricultural sector from farm to mouth by

1. Further limiting preharvest crop losses to pests and diseases, which currently reduce global yields by an estimated 42 percent;
2. Increasing fertilizer use;
3. Liming acidic soils;
4. Adapting high-yield varieties to specific locations around the world, although many scientists believe that opportunities to further increase yields through conventional breeding techniques are almost tapped out; and
5. Reducing postharvest and end-use losses, which are estimated at about 47 percent worldwide.

Moreover, precision farming could help reduce chemical and water use without reducing yields, which would reduce many of the adverse effects of modern agriculture. Productivity improvements could be made much more rapidly and more surely with the aid of biotechnology. Biotechnology could more easily and quickly reduce current gaps between average yields and yield ceilings, and yield ceilings and the theoretical maximum yield, as well as push up the theoretical maximum yield. This begs the question as to whether such increases in productivity will also increase environmental costs and whether the productivity increases are sustainable in the long run. That issue will be discussed in greater detail later.

If the widespread adoption of biotechnology allowed agricultural productivity to increase from 1 percent to 1.5 percent per year, cropland could actually be *reduced* by 98 million hectares rather than increased by 325 million hectares (relative to 1997 levels), and at the same time the increased food demand of a larger and richer population could be met. That corresponds to a net increase in productivity of 30 percent in 2050 attributable to biotechnology alone. And if productivity is increased 2 percent per year, by 2050 at least 422 million hectares of cropland currently under plow could be returned to nature or made available for other human uses. That corresponds to a net improvement in productivity by 2050 of 69 percent due to biotechnology.

Several biotechnologically engineered crops, currently in various stages between research and commercialization, could increase yields and, more importantly, put more food on the table per unit of land and water diverted to agriculture. Such crops, which could be particularly useful in developing nations, include

- *Cereals that are tolerant of poor climatic and soil conditions*—specifically, cereals that can tolerate aluminum (they can grow in acidic soils), drought, high salinity levels, submergence, chilling, and freezing. The ability to grow crops in such conditions could be critical for developing countries: 43 percent of tropical soils are acidic; more cropland is lost to high salinity than is gained through forest clearance; and salinity has rendered one-third of the world's irrigated land unsuitable for growing crops. Moreover, if the world warms, the ability to tolerate drought, high salinity, submergence, and acidity could be especially important in achieving global food security. In Kasuga et al.'s experiments, 96.2 percent of genetically modified plants survived freezing compared with 9.5 percent of conventional plants. The corresponding numbers for drought were 76.7 percent vs. 1.8 percent and, for salinity stress, 78.6 percent vs. 17.9 percent.

- *Rice that combines the best traits of the African and Asian varieties*— the ability of African rice to shade out weeds when young and the high yield capacity of the Asian variety. In addition, the GM variety is highly resistant to drought, pests, and diseases—a trait that could be particularly useful for Africa because its increases in rice yields have so far lagged behind those of the rest of the world. That lag is one reason why malnourishment in Sub-Saharan Africa has increased in the past several decades, in contrast to improved trends elsewhere.
- *Rice that can close stomata more readily.* Use of such rice should increase the efficiency of water use and net photosynthetic efficiency. Both aspects will be useful under dry conditions—conditions that may become more prevalent in some areas as a result of global warming.
- *Rice with the alternative C4 pathway for photosynthesis.* This trait could be especially useful if there is significant warming because the C4 pathway is more efficient at higher temperatures. In addition, efforts are under way to try to reengineer RuBisCO—an enzyme critical to all photosynthesis—by using RuBisCO from red algae, which is a far more efficient catalyst for photosynthesis than that found in crops. *Maize, rice, and sorghum that are resistant to* Striga, *a parasitic weed that could decimate yields in Sub-Saharan Africa. Rice that has the ability to fix nitrogen.* Nitrogen, an essential plant nutrient, limits plant growth in much of the world. While ample supplies of nitrogen are available in the air and in the soil, little of it can be used directly by plants because it is not "fixed" in a form that plants can use directly. Hence the need for fertilizers

containing "fixed" nitrogen. In the beginning of the 20th century, with resources of guano—a rich source of fixed nitrogen—running out, it seemed that global food production— and hence the world's population—was about to reach its limit. However, these fears were allayed with Fritz Haber's 1909 invention of a process to synthesize ammonia—a nitrogen fertilizer—from nitrogen in the air and hydrogen. For this signal achievement, which not only foreshadowed—but also made possible—the green revolution in the last third of the 20th century, Haber received the 1918 Nobel Prize for Chemistry. Nitrogen-fixing genetically modified crops would reduce humanity's dependence on synthetic nitrogen fertilizers, and the several environmental problems associated with their use, and overuse. They would also reduce costs to farmers and consumers alike.

- *Rice and maize with enhanced uptake of phosphorus and nitrogen*. Notably, these two crops account for 20 percent of global cropland.
- *Rice, maize, potato, sweet potato, and papaya that are resistant to insects, nematodes, bacteria, viruses, and fungi.* For instance, papaya, which had been ravaged in Hawaii by the papaya ringspot virus, has now made a comeback due to a bioengineered variety resistant to that virus.
- *Cassava, a staple in much of Africa, that is resistant to the cassava mosaic virus and including a gene with an enzyme (replicase) capable of disrupting the life cycles of a number of other viruses.* It is claimed that this GM cassava could increase yields 10-fold. In addition, because cassava contains natural substances that can be converted to cyanide, it must be carefully prepared before consumption. Work is proceeding on producing a GM cassava that would be less toxic.
- *Spoilage-prone fruits bioengineered for delayed ripening to increase their shelf life and reduce post-harvest losses.* Such fruits include bananas and plantains, important

sources of food for many African nations, and, in the United States, melons, strawberries, and raspberries.

- *Crops bioengineered to reduce the likelihood of their seed pods shattering*—which reduces yields of crops such as wheat, rice, and canola. It is estimated that bioengineering could increase canola yields, for instance, by 25 percent to 100 percent.
- *High-lysine maize and soybeans, maize with high oil and energy content, and forage crops with lower lignin content*—which ought to improve livestock feed and reduce the overall demand for land needed for livestock.

If the methods and genes used to bioengineer those crops can be successfully adapted and transferred to other vegetables, tubers, fruits, and even trees, it would help reduce the amount of land and water needed in the future to feed, clothe, and shelter human populations, and it would free up those resources for the rest of nature.

Use of the GM crops described above, by increasing crop yields and reducing the amount of cultivated land, would also reduce the area subject to soil erosion from agricultural practices. That, in turn, would limit associated environmental effects on water bodies and aquatic species and reduce the loss of carbon sinks and stores into the atmosphere. Furthermore, many of the same GM crops could also directly reduce nutrients and pesticides released into the environment. These bioengineered crops include

- *Nitrogen-fixing rice and rice and maize bioengineered with the ability to increase uptake of phosphorus and nitrogen from the soil.* In Europe and the United States, only 18 percent of the nitrogen and 30 percent of the phosphorus in fertilizers are incorporated into crops, between 10 and 80 percent of the nitrogen and 15 percent of the phosphorus end up in aquatic ecosystems, and much of the remainder accumulates in the soil, to be later eroded into aquatic systems. Nitrogen-fixing crops would reduce reliance on fertilizers and thereby reduce ground and surface water pollution, risks of chemical spills, and atmospheric emissions of nitrous oxide (N_2O), a greenhouse gas that, pound for pound over a 100-year period, is 310 times more potent a greenhouse gas than CO_2.

- *Crops resistant to viruses, weeds, and other pests*—for example, *Striga*-resistant maize, rice, and sorghum. Examples also include various *Bt* crops containing genes from the *Bacillus thuringiensis* bacterium, which has been used in insecticide spray for four decades. Carpenter and Gianessi estimate that in 1999 the planting of *Bt* cotton on 32 percent of U.S. cotton areage reduced chemical pesticide use by over 2.7 million pounds, increased yields by 260 million pounds, and netted farmers $99 million. Although the use of *Bt* corn—which is planted on 26 percent of the acreage used to grow corn in the United States—resulted in a drop in farmers' revenues of $35 million, pesticide use dropped by 1 million acre-treatments, and production increased by 66 million bushels, equivalent to the output of half a million acres. Developing countries also can reduce the use of pesticides by using pest-resistant crops. For example, in India—the world's third largest producer of cotton—cotton occupies only 5 percent of the land yet cotton farmers buy about 50 percent of all pesticides used in the country. In 1998, the devastation caused by pests reportedly contributed to 500 suicides among Indian cotton farmers whose crops had failed. Field trials of *Bt* cotton at 30 locations in India show a 14 to 38 percent increase in yield despite suspension of all spraying.
- *Low-phytic-acid corn and soybean and phytase feed*—which help livestock better digest and absorb phosphorus. This would reduce phosphorus in animal waste and decrease runoff into streams, lakes, and other water bodies, mitigating one of the major sources of excess nutrients in the environment. It would also reduce the

need for inorganic phosphorus supplements in feed.

Crops tolerant of various herbicides—so that those herbicides can be used to kill weeds but not the crop itself. Herbicide-tolerant crops are among the most common applications of biotechnology today. One commercially available example is "Roundup Ready" soybean, which is engineered to be tolerant to glyphosate. Such crops could help reduce the amount, toxicity, and/or persistence of pesticides. Planting of such crops seems to have reduced application of more hazardous and longer-lasting herbicides (e.g., acetochlor), although overall herbicide use may have increased. Gianessi and Carpenter estimate that U.S. farmers who planted Roundup Ready soybean (which was planted on 47 percent of soybean acreage) lowered weed control costs by $216 million (equivalent to 19 million acre-treatments). Second, such crops would also increase yields while facilitating no-till cultivation, which, by stemming soil erosion, protects future agricultural productivity. Moreover, erosion can be particularly damaging to the environment because the eroded particles can transport fertilizers and pesticides into aquatic systems and into the atmosphere. Finally, as noted, soil erosion releases stored carbon into the atmosphere.

Bioengineered crops can also be engineered to directly clean up environmental problems. For instance, GM plants can be developed that selectively absorb various metals and metal complexes such as aluminum, copper, and cadmium from contaminated soils. Such plants could, for example, detoxify methyl mercury in soils, removing it from the food chain.

Researchers have also genetically modified aspen trees to produce 50 percent less lignin and 15 percent more cellulose. Lignin, a component of all wood, must be chemically separated from cellulose to make the pulp used in paper production. The GM tree has half the normal lignin-to-cellulose ratio of about 1:2. Overall, 15 percent more pulp may be produced from the same amount of wood.

Moreover, the GM trees are 25–30 percent taller. Thus, the amount of land, chemicals, and energy required to make a given quantity of paper should to be reduced substantially, as should the environmental impacts at every stage, from tree farming to paper production. Other potential applications of biotechnology that could reduce environmental impacts include production of biodegradable plastics using oilseed rape and colored cotton (which could reduce reliance on synthetic dyes).

Health Benefits of Bioengineered Crops

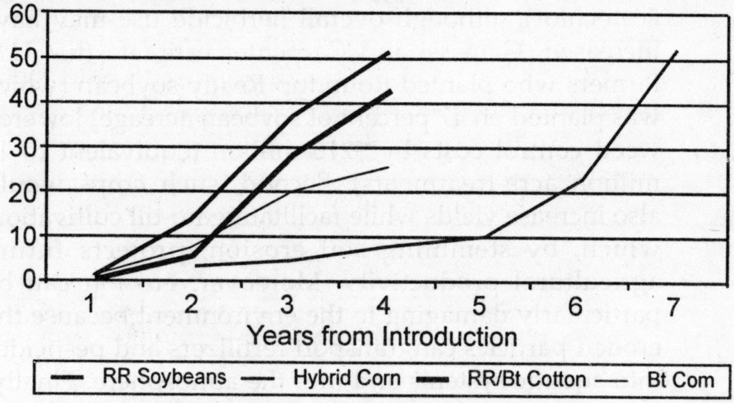

Fig. Adoption of Bioengineered Crops

Having sufficient quantities of food is often the first step to a healthy society. The increase in food supplies per capita during the past half-century is a major reason for the worldwide improvement in human health during that period. Between 1961 and 1997, food supplies per capita increased 23 percent. Thus, despite a 40 percent increase in population between 1969–71 and 1994–96, chronic undernourishment in developing countries dropped from 35 percent to 19 percent of their population, which in turn helped lower global infant mortality rates from 156 to 57 per 1,000 live births, increase life expectancies from 46.5 to 65.7 years, and enable the average person to live a more fulfilling and productive life.

Despite unprecedented progress during the 20th century, billions of people still suffer from undernourishment, malnutrition, and other ailments attributable, in whole or in part, to insufficient food or poor nutrition. Reducing those numbers over the next half-century—while also reducing pressures on biodiversity despite anticipated population increases of 1.3 to 4.7 billion—requires increasing the quantity of food produced per unit of land and water. As discussed above, GM crops could help in that struggle. But increasing the quantity of food is not enough. Improving the nutritional quality of food is just as important. The diets of over half the world's population are deficient in iron, vitamin A, and other micronutrients. Such deficiencies can cause disease, if not death. About 2 billion people do not have enough iron in their diet and are thus susceptible to anemia.

Another 260 million suffer from vitamin A deficiency, which causes clinical xerophthalmia, a condition that if untreated, may lead to blindness, especially in children. Vitamin A is also crucial for effective functioning of the immune system. Through the cumulative effect of these deficiencies, malnutrition in 1995 was responsible for 6.6 million or 54 percent of deaths worldwide in children under five, stunted growth in another 200 million children, and clinical xerophthalmia in about 2.7 million people.

In addition to ensuring that adequate quantities of food are available, bioengineering could also help reduce many of those micronutrient deficiencies. For instance, Swiss scientists have developed "golden rice"—rich in betacarotene, a precursor to vitamin A—and crossed it with another bioengineered strain rich in iron and in cysteine, which allows iron to be absorbed in the digestive tract. Such rice would help reduce vitamin A and iron deficiency–related deaths and diseases in the developing world. Iron-fortified rice, whether golden or not, would also reduce the need for meat, one of the primary sources of dietary iron. As a result, overall demand for livestock feed—and the land, water, and other factors needed to produce that feed—might be reduced.

Scientists are also working on using bananas and other fruits as vehicles to deliver vaccines against the Norwalk virus, *E. coli*, hepatitis B, and cholera, eventually leading to low-cost, efficient immunization of whole populations against common diseases, with broader coverage than is likely with conventional needle immunization.

Bioengineered crops can also help battle the so-called "diseases of affluence": ischemic heart disease, hypertension, and cancer. In 1998, according to the World Health Organization, those diseases accounted for 4.8 million or 60 percent of the total deaths in high-income countries and 14.9 million or 32 percent of deaths in low- and middle-income countries. There are several GM crops that can help reduce this toll. For instance, genetically enhanced soybeans that are lower in saturated fats are already on the market.

The International Food Information Council notes that biotechnology could also make soybean, canola, and other oils and their products, such as margarine and shortenings, more healthful. Bioengineering could also produce peanuts with improved protein balance; tomatoes with increased antioxidant content; potatoes with higher starch than conventional potatoes, which ought to reduce the amount of oil absorbed during the processing of foods like French fries or potato chips; fruits and vegetables fortified with or containing higher levels of vitamins such as C and E; and higher-protein rice, containing genes transferred from pea plants.

Moreover, levels of mycotoxins, which apparently increase with insect damage in crops, are lower in *Bt* corn. Some mycotoxins, such as fumonisin, can be fatal to horses and pigs and may be human carcinogens. Morton argues that GM *Bt* food crops are safer than conventional crops sprayed with *Bt* because the sprays contain several toxins that could affect both insects and mammals, whereas the GM variety contains a single toxin known to be harmful to insects but not to mammals. Thus, *Bt* corn, whether used as food for humans or feed for livestock, may be safer and healthier than conventional corn.

GM plants may be further able to save life and limb if they can be successfully engineered to biodegrade explosives around land mine and abandoned munitions sites. Finally, to the extent that pest-resistant GM plants can reduce the amount, toxicity, and persistence of pesticides used in agriculture (by themselves or as parts of integrated pest management systems), it would reduce accidental poisonings and other harmful health effects on farm workers. For instance, there have apparently have been instances of food poisoning and human infections from *Bt* sprays but none (so far) from *Bt* crops.

Adverse Environmental Consequences

The major environmental concerns regarding GM crops are those related to crops that are designed to be resistant to pests or tolerant of herbicides. One potential risk is that target pests will become resistant to toxins produced by pest—resistant GM crops, such as *Bt* corn or *Bt* cotton. Although that is a possibility even if *Bt* is delivered via conventional sprays on non-GM plants, it is argued that it is of greater concern with *Bt* plants because with conventional spraying target pests are exposed to *Bt* toxins for only brief periods, whereas currently available *Bt* crops produce toxins throughout the growing season, which could increase the chances of developing *Bt*-resistant pests.

Moreover, some labora tory studies suggest that target pests may develop resistance more rapidly than had previously been thought possible. However, subsequent studies from Arizona, Mississippi, and Australia indicate that contrary to these prognostications, bollworm, for instance, did not increase its resistance to *Bt* toxin produced by a GM *Bt* cotton. It has also been argued that the only known insect resistance to *Bt* is caused by *Bt* sprays. This has been attributed to the adaptation of conventional strategies (developed to deter pest resistance due to conventional pesticide) to GM crops. Such strategies include ensuring that plants deliver high doses of *Bt* while simultaneously maintaining refuges for non-*Bt* crops so that pest populations remain susceptible to *Bt*.

In fact, the Environmental Protection Agency has established the requirement that *Bt* corn farmers plant 20 percent of their land in non-*Bt* corn, as refuges. For *Bt* corn grown in cotton areas, farmers must plant at least 50 percent non- *Bt* corn. EPA also requires expanded monitoring to detect any potential resistance. Other strategies to delay development of pesticide resistance include crop rotation; developing crops with more than one toxin gene acting on separate molecular targets; and inserting the bioengineered gene into the chloroplast, which ought to express *Bt* toxin at higher levels.

Notably, farmers have an economic stake in implementing such adaptive strategies so that their crop losses to pests are kept in check in the long, and the short, term.

Another source of risk is that *Bt* from pest-resistant plants could harm, if not kill, nontarget species. That could happen if, for instance, *Bt*-laden pollen were to drift away from the field or if the toxin were to leak through the roots and be consumed by nontarget organisms susceptible to the *Bt* toxin.

However, a recent study suggests that such root leakage is quite unlikely to either kill non-target pests or, for that matter, cause *Bt* to accumulate in non-*Bt* crops grown subsequently on the same soil. But in a laboratory study that captured headlines around the world, and stoked the flames of the anti-GM cause, Losey et al. indicated a 44 percent mortality rate for monarch butterfly larvae fed on milkweed dusted with *Bt* corn pollen compared with zero for the control case (which used milkweed dusted with ordinary pollen). In a more sophisticated study, Jesse and Obrycki found a 20 percent mortality rate for monarch butterfly larvae fed in the laboratory with leaves exposed to pollen in or near a field of *Bt* plants.

However, whether—and the extent to which—the monarch butterfly population would be affected in the real world is a matter of debate. One study suggests that under a worst-case scenario as much as 7 percent of the North American population (estimated at 100 million) may die, although the real-world effect would probably be smaller. Some have also argued that the major threat to monarchs is the habitat loss in their wintering grounds in Mexico, which is a result of pressure from a growing population in need of land.

Notably, in a recent analysis, EPA concluded that based on their examination, "the weight of evidence" indicates "no hazard to wildlife from the continued registration of *Bt* crops." The Agency also concluded that continued cultivation of *Bt* corn is unlikely to "cause harmful widespread effects to monarch butterflies at this time." It also noted that the only endangered species of concern are in the lepidoptera and coleoptera group (i.e., butterflies, moths, and beetles), but the majority of those species have very restricted habitat range and do not feed in, or close to the *Bt* crop planting areas. Perhaps more importantly, the inadvertent effects of *Bt* crops due to pollen dispersal or root leakage could be virtually eliminated by bioengineering genes into the chloroplast rather than into nuclear DNA.

Bt could also enter the food chain through root leakage or if predators prey on target pests. For instance, studies have shown that green lacewing larvae, a beneficial insect, that ate maize borers fed with *Bt* maize were more likely to die. But the real-world significance of this has also been disputed based on the long history of *Bt* spraying on crops and other studies that have shown beneficial insects essentially unharmed by such spraying, particularly under field conditions.

There is also a concern that bioengineered genes from herbicide- or pest-tolerant crops might escape into wild relatives, leading to "genetic pollution" and creating "superweeds." That would have an ad verse economic impact on farmers, reducing crop yields and detracting from the very justification for using such GM crops. Clearly, the farmer has a substantial incentive for preventing weeds from acquiring herbicide tolerance and, if that fails, to keep such weeds in check.

Gene escape is possible if sexually compatible wild relatives are found near fields planted with GM crops, as is the case in the United States for sorghum, oats, rice, canola, sugar beets, carrots, alfalfa, sunflowers, and radishes. However, the most common GM crops—soybeans and corn—have no wild U.S. relatives. As the Royal Society pointed out in its assessment of the issue, centuries of conventional breeding have rendered a number of important crops, such as, maize and wheat, "ecologically incompetent" in many areas. It also noted that despite the use of conventionally bred herbicide-tolerant plants, there has been no upsurge in problems related to herbicide-tolerant weeds.

Although these theoretical arguments by themselves do not guarantee safety, they seem confirmed by Crawley et al.'s 10-year-long British study of four different herbicide-tolerant or pest-resistant GM crops (oilseed rape, corn, sugar beet, and potato) and their conventional counterparts grown in 12 different habitats. That study indicated that within four years all plots of rape, corn, and beet had died out naturally. Only one plot of potatoes survived the 10th year, but that was a non-GM variety. In other words, GM plants were no more invasive or persistent in the wild than their conventional counterparts. And had any herbicide-tolerant or pest-resistant weeds begun to spread, available crop management techniques (such as another herbicide) could have been used to control them.

The Crawley et al. study also provides reassurance with respect to another potential environmental concern: that herbicide-tolerant or pest-resistant "superweeds" could invade natural ecosystems. The study confirms that such GM plants do not have a competitive advantage in a natural system unless that system is treated with the herbicide in question. But if it were so treated, would it still qualify as a natural system? Moreover, if it had to be treated, another herbicide to which the so-called superweed is not resistant could be used. On the other hand, if the area is not treated with the herbicide in question, what difference does it make to the ecosystem whether the weed is tolerant? And what is the significance of "genetic pollution" with respect to ecosystem function and biodiversity? Would gene escape affect ecosystem function negatively? Does gene escape diminish or expand biodiversity?

To bring this issue into focus, consider the case of human beings. If an Indian from Calcutta came to Washington, D.C., and had an offspring with a native-born American, would that not, as the term has been used, be considered "genetic pollution"? (Not very long ago, xenophobes labeled that miscegenation.) Does that diminish or expand biological diversity? Is such genetic pollution acceptable for human beings but not for other species? But if the answer varies with the species, it raises questions about the validity of the notion that gene escape can be equated to pollution—genetic or otherwise.

In addition, genes may escape from GM crops to non-GM crops of the same species. If that were to occur, it would be unpopular with organic farmers, who are afraid it might "adulterate" their produce, and with producers and farmers of GM seeds, who are not eager to have someone else profit from their investments. Crawley et al.'s study is consistent with the Royal Society's prognosis that because more crops (including corn, sorghum, sugar beets, and sunflowers) are now grown from hybrid seeds, that provides a measure of built-in security against such gene transfers. Moreover, the chances of such gene escape can be further reduced by maintaining a buffer between the two crops.

Of course, gene escape could be limited with greater certainty if the GM plant were engineered to be sterile or were prevented from germinating by using, for instance, "terminator technology." An alternative approach would be to insert the gene into the chloroplast, which would preclude spread through pollen or fruit and prevent root leakage.

Finally, there is a concern that in the quest to expand yields GM plants will work too well in eliminating pests and weeds, and that this will lead to a further simplification of agricultural ecosystems and a further decrease in biodiversity. That concern, in conjunction with the other noted environmental concerns, needs to be weighed against the cumulative biodiversity and other environmental bene fits of reduced conversion of habitat to cropland, and decreased use of chemical inputs.

A major health concern is that the new genes inserted into GM plants could be incorporated into a consumer's genetic makeup. However, there is no evidence that any genes have ever been transferred to human beings through food or drink despite the fact that plant and animal DNA has always been a part of the daily human diet. In fact, an estimated 4 percent of the human diet is composed of DNA, and an average adult Briton consumes 150,000 km of DNA in an average meal. Moreover, it is unclear whether consuming, for instance, beans that have been genetically modified with genes from a pig would pose a greater risk to public health than consuming a dish of non-GM pork-and-beans.

Another concern is that genes transferred from foods to which many people are allergic could trigger allergies in unsuspecting consumers of such GM crops. Between 1 and 3 percent of the adults and between 5 and 8 percent of the children in the United States suffer from food allergies. Each year food allergies cause 135 fatalities and 2,500 emergency room visits.

The concern regarding allergic reactions to GM foods can be traced to precommercialization tests conducted by Pioneer Hi-Bred, which showed that a soybean that had been bioengineered to boost its nutritional quality by using a gene from the Brazil nut was, in fact, allergenic. Although this example shows that GM foods can be tested prior to commercialization for their allergic potential, opponents of GM foods have used this as an argument against bioengineered crops. Several databases of known allergens could be used to help identify problematic GM products before they are developed. In fact, because bioengineering allows more precise manipulation of genes than does conventional plant breeding, it could be used to render allergenic crops nonallergenic.

Yet another potential negative effect on public health is that antibiotic resistant "marker" genes, which are used to identify whether a gene has been successfully incorporated into a plant, could, through consumption of the antibiotic gene by humans, accelerate the trend toward antibiotic-resistant diseases. However, by comparison with the threat posed by the use of antibiotics in feed for livestock and their overuse as human medicines, the increased risk due to such markers is slight.

Applying the Precautionary Principle

The above discussion indicates that there are risks associated with both the use and the nonuse of GM crops. Ideally, each criterion should be applied, one at a time, to the consequences of GM crop use (or nonuse) for human mortality, public health, and the environment. However, because there are variations in the severities, certainties, and magnitudes associated with the various competing costs and benefits of each of these sets of consequences, one may have to apply several criteria simultaneously.

Public Health Consequences

Population could increase 50 percent between 1998 and 2050 (from 5.9 billion to 8.9 billion, according to the best UN estimate). Hence, by 2050, one ought to expect that undernourishment, malnutrition, and their consequences—death and disease—would also increase by 50 percent worldwide, if the global food supply increases by a like amount and all else remains equal. Thus, which provides estimates for the prevalence of such problems in the mid- to late-1990s, unless food production outstrips population growth significantly over the next half century, billions in the developing world may suffer annually from undernourishment, hundreds of millions from stunted growth, and millions may die from malnutrition.

Based on the sheer magnitude of people at risk of hunger and malnutrition, and the degree of certainty regarding the public health consequences, one can state with confidence that limiting GM crops will, by limiting'the rate at which food production can expand, almost certainly increase death and disease, particularly among the world's poor.

GM crops could also reduce or postpone deaths due to diseases of affluence. Although the probability that that might occur is lower than the probability of reducing deaths from hunger and malnutrition, the expected number of deaths postponed could run into the millions. A 10 percent decrease, for instance, in today's 15 million annual deaths from cancer and ischemic and cerebrovascular diseases translates into 1.5 million lives saved each year. And those numbers could increase in the future as populations increase and become older.

By contrast, the negative public health consequences of ingesting GM foods are speculative (e.g., the effects due to ingesting transgenes), relatively minor in magnitude (e.g., a potential increase in antibiotic resistance), or both (e.g., increased incidence of allergic reactions). Moreover, it is possible to contain, if not eliminate, the effects of even those impacts. As noted previously, the likelihood of allergic reactions can be reduced by checking various databases of known allergens prior to developing a GM crop and by testing food from such crops prior to commercialization.

Progress is also being made with respect to the risk of increasing antibiotic resistance. Novartis has developed a sugar-based alternative to antibiotic-resistant marker genes that has been used to develop about a dozen GM crops, including maize, wheat, rice, sugar beets, oilseed rape, cotton, and sunflowers. With additional research, it ought to be possible to devise alternative marker genes for other crops or develop practical methods to remove or repress antibiotic-resistant marker genes.

Thus, based on the *uncertainty, expectation value,* and *adaptation criteria* applied either singly or in conjunction, use of genetic modification must be favored over nonuse. Hence, the precautionary principle *requires* that we continue to research, develop, and commercialize (with appropriate safeguards, of course) those GM crops that would increase food production and generally improve nutrition and health, especially in the developing world.

Some have argued that many developed countries are "awash in surplus food" and have no need to boost food production. However, that argument ignores the fact that reducing those surpluses would be almost as harmful to public health in developing countries as curtailing those countries' own food production. At present, net cereal imports of the developing countries exceed 10 percent of their production.

Without trade (and aid) that moves the surplus production in developed countries voluntarily to developing countries suffering from food deficits, food supplies in developing countries would be lower, food prices would be steeper, undernourishment and malnutrition would be higher, and associated health problems, such as illness and premature mortality, would be greater. As already noted, developing countries' food deficits are expected to increase in the future because of high population growth rates and possibly further worsened by global warming. Therefore, the food surpluses of developed countries will be at least as critical for future food security in developing countries as they are today.

The preceding argument against GM crops implicitly assumes that such crops will provide little or no public health benefits to the inhabitants of developed countries. As noted above, GM crops are also being engineered to improve nutrition in order to combat diseases of affluence afflicting populations in developed, as well as developing, nations. Such diseases, which are major causes of premature death worldwide, kill about 4.8 million annually in the developed countries.

If the use of GM crops were to produce even a small reduction in these numbers it would translate into relatively large declines in the death tolls of developed countries. The health benefits of "golden rice," for instance, do not have to be confined to developing countries; developed countries, too, could avail themselves of its benefits. Thus even for developed countries, the magnitude and certainty of the potential public health benefits of GM crops far outweigh the speculative health consequences of ingesting GM foods.

Another argument against using GM foods to increase food production is that, because there is no shortage of food in the world today and the problem of hunger and malnutrition is rooted in poor distribution and unequal access to food because of poverty, it is unnecessary to increase food production; therefore, there is no compelling need for biotechnology. Significantly, that argument tacitly acknowledges that GM crops would boost production (and productivity). The rest of the argument, however, is flawed.

First, while unequal access is a perennial problem—one that continues to persist despite the successes of conventional agriculture— the argument misses the point. The case for bioengineered crops is not that it is the one and only solution for solving hunger and malnutrition. Increasing food production and improving access are not mutually exclusive. It should be sufficient that GM crops can contribute to the solution and that they are among the most efficient solutions for the problem.

Conventional agriculture has been relatively successful in improving global access to food. The evidence lies in the fact that hunger and malnourishment have declined substantially in the past few decades despite a substantial increase in population. The principal reasons for these improvements are production increases that outstripped population growth and brought food prices down worldwide; economic growth, which made food more affordable to all, particularly in developing countries; investments in infrastructure, which enable rapid and efficient distribution of agricultural products; and greater democratization, which increases the political accountability of rulers to those ruled.

Although GM crops (in comparison with conventional crops) are unlikely to directly increase democratization or economic growth, they can indirectly boost the latter. Moreover, as the anti-GM crop argument (that there is sufficient food) seems to recognize, GM crops could increase productivity more rapidly and by larger amounts, which would further increase food availability and reduce costs to consumers. In addition, one of the problems contributing to poor distribution is spoilage of crops before they are consumed. As noted, various GM crops could increase shelf life and reduce spoilage and wastage. GM crops cannot guarantee equal access, but they can improve access for the poorer segments of society more rapidly than can conventional crops.

Second, if the argument that there is sufficient food is truly a compelling one against GM crops, it should be equally valid for increases in production using conventional technologies. If that is indeed the case, perhaps developing countries like India and Bangladesh should forego increasing agricultural productivity altogether and focus only on improving access and distribution. For obvious reasons, no one makes this argument.

Environmental Consequences

Regardless of the level of demand, limiting GM crops would lower crop and forest yields per unit of land and water used. Reductions in the amount of land and water available for the rest of nature would be further aggravated by the fact that the price of land and water, relative to other goods, would rise, as would the opportunity costs for those resources. As a result, the socioeconomic costs of setting aside land or water for conservation and preservation of nature would increase, further inhibiting *in situ* conservation of species and biodiversity, which is one of the major goals of the Convention on Biological Diversity, beyond the goal of merely conserving biodiversity itself.

If bioengineering succeeds in improving the protein and micronutrient content of vegetables, fruits, and grains, it might persuade many more people to adopt and, more importantly, persevere with vegetarian diets, thereby reducing the additional demand meat-eating places on land and water. In addition, giving up GM crops will, more likely than not, further increase the pressures on biodiversity caused by excess nutrients, pesticides, and soil erosion. Finally, reduced conversion of habitat and forest to cropland and timberland, coupled with reductions in soil erosion due to increased no-till cultivation, would further limit deterioration of water quality and losses of carbon reservoirs and sinks.

Arrayed against these benefits to ecosystems, biodiversity, and carbon stores and sinks are the environmental costs of limiting pestresistant and herbicide-tolerant GM crops *minus* the environmental costs of conventional farming practices. These costs include a potential increase in the diversity of the flora and fauna associated with or in the immediate vicinity of GM crops if they are more effective in reducing nontarget pests and weeds than conventional farming prac tices, and the possible consequences of gene escape to weeds and non-GM crops.

Therefore, with respect to the environmental consequences of the use or nonuse of GM crops, one must conclude, based on the *uncertainty* and *expectation value criteria*, that the precautionary principle requires the cultivation of GM crops. On balance, GM crops should conserve the planet's habitat, biodiversity, and carbon stores and sinks, provided due caution is exercised, particularly with respect to herbicide-tolerant and pest-resistant GM crops.

It may be argued that although gene escape to "natural" ecosystems might be a low-probability event, it may cause irreversible harm to the environment; thus, under the *irreversibility criterion*, GM crops ought to be banned. However, the increased habitat loss and land conversion that would result from such a ban may be at least as irreversible, particularly if they lead to the extinction of species.

It is worth noting that the precautionary principle supports using terminator-type technology because that would minimize the possibility of gene transfer to weeds and non-GM plants without diminishing any of the public health or environmental benefits of GM crops. It is notable that some of the same groups that profess environmental concerns about genetic pollution have subjected terminator technology to unbridled criticism.

In these groups' policy calculus, the potential environmental costs of GM crops are outweighed by the presumption that farmers will suffer economically because they will not be able to propagate GM crops from sterile seeds, and by antipathy toward multinationals' profits. It is debatable whether putting antipathy to profits ahead of public health or the environment is any more commendable than putting profits ahead of them. It certainly does not advance either human or environmental well-being.

The Goals of the Convention on Biological Diversity

The above analysis indicates that banning GM crops would more likely than not magnify threats to biodiversity and *in situ* conservation. But the preamble to the Convention on Biological Diversity (CBD) states that "it is vital to anticipate, prevent and attack the causes of significant reduction or loss of biological diversity at source" and that "the fundamental requirement for the conservation of biological diversity is the in-situ conservation of ecosystems and natural habitats and the maintenance and recovery of viable populations of species in their natural surroundings". Article 1 of the CBD also identifies as the first of its various objectives "the conservation of biological diversity." Thus, a ban on GM crops would be directly counter to the Convention's raison d'être.

Banning GM crops would also contradict the letter of the CBD. Article 8(d), which addresses *in situ* conservation, requires that "each contracting party shall, as far as possible and as appropriate ... promote the protection of ecosystems, natural habitats and the maintenance of viable populations of species in natural surroundings". A GM ban would also make it harder to satisfy the requirements of Article 8(a), which requires contracting parties to "establish a system of protected areas or areas where special measures need to be taken to conserve biological diversity", because, as noted above, there would be less land and water available for *in situ* conservation, and what would be available would be socially and economically costlier to obtain and maintain.

The Goals of the Cartagena Biosafety Protocol

Both the objective and scope of the Cartagena Biosafety Protocol are specifically limited to GM organisms (GMOs) "that may have adverse effects on the conservation and sustainable use of biological diversity, taking also into account risks to human health". But we have seen that in the aggregate, by contrast with conventional crops, GM crops not only are unlikely to have adverse impacts on the environment, conservation, sustainable use, and human health but might result in net benefits. Thus, it can be argued that the case for a general ban on GM crops under the Protocol is, at best, weak. And it looks even weaker when one considers the sources of the Protocol's authorities.

As noted in its preamble and Article 1, the Protocol derives its authority regarding the regulation of the transfer, handling, and use of GMOs largely from Principle 15 of the Rio Declaration, and Articles 17, 8(g), 19(3), and 19(4) of the CBD. Let's examine whether a general ban on GM crops would be supported under each of these sources of authority:

- Principle 15 of the Rio Declaration is a statement of the precautionary principle. But in previous sections we saw that the precautionary principle does not support a broad ban on GM crops.
- Article 17 of the CBD addresses only the facilitation of an exchange of information, so it cannot be used to justify a general ban on GM crops either.
- Article 8(g) of the CBD requires that contracting parties shall "establish or maintain means to regulate, manage or control the risks associated with the use and release of living modified organisms resulting from biotechnology which are likely to have adverse environmental impacts that could affect the conservation and sustainable use of biological diversity, taking also into account the risks to human health." This language parallels that in Articles 1 and 4 of the Protocol. The same rationale enunciated in the previous section applies to why it would be inappropriate to use that language to ban GM crops—namely, that GM crops in the aggregate are likely to have positive rather than adverse impact on conservation and sustainable use of biodiversity, and on human health.
- The same rationale explains why Article 19(3) of the CBD cannot be used to justify a GM crop ban. This Article requires that "the Parties shall consider the need for and modalities of a protocol setting out appropriate procedures, including, in particular, advance informed agreement, in the field of the safe transfer, handling and use of any living modified organism resulting from biotechnology that may have adverse effect on the conservation and sustainable use of biological diversity."

- Article 19(4) of the CBD operates only through Article 19(3). But, because Article 19(3) cannot justify a GM crop ban, neither can Article 19(4).

A counterargument could be made that the appropriate test to determine whether GM crops as a class should be subject to the Protocol's requirements would be to evaluate whether it may have *any*— and not just *aggregate*—adverse impact on the environment, conservation, sustainable use, and human health. However, as the *any* in Article 19(3) of the CBD indicates, the CBD's negotiators seem to have been cognizant of that word's significance: *any* appears 43 times in the Protocol (and 66 times in the Convention).

More importantly, had the CBD explicitly said *any* adverse impact, and had that then been used to impose a blanket ban on GM crops, it would have been counterproductive—and inconsistent with the goals of the Convention and the Protocol. The precautionary principle has often been cited to justify a prohibition on GM crops. However, that justification is based on a selective application of the principle to a limited set of consequences of such a policy.

Specifically, the justification takes credit for the potential public health and environmental benefits of a ban on GM crops but refuses to accept responsibility for any concomitant risks that might be created or prolonged. By comparison with conventional crops, GM crops would, in fact, increase the quantity and nutritional quality of food supplies and thereby improve public health by reducing mortality and morbidity rates worldwide. In addition, cultivation of GM rather than conventional crops would, by increasing productivity, reduce the amount of land and water that would otherwise have to be diverted to mankind's needs.

GM crops could also reduce the environmental damage from the use of synthetic fertilizers and pesticides and from soil erosion. In comparison with conventional crops, GM crops would be more protective of habitat, biological diversity, air and water quality, carbon stores and sinks, and the world's atmosphere. A ban on GM crops—whether accomplished directly by invoking the precautionary principle or indirectly via application(s) of the Biosafety Protocol—is likely to aggravate threats to biodiversity and further increase the already considerable hurdles facing *in situ* conservation.

Therefore, a ban would be counterproductive and contravene the spirit and the letter of the CBD. The precautionary principle—properly applied, with a more comprehensive consideration of the public health and environmental consequences of a ban—argues instead for a sustained effort to research, develop, and commercialize GM crops, provided reasonable caution is exercised during testing and commercialization of the crops. In this context, a "reasonable" precaution is one whose public health benefits are not negated by harmful reductions or delays in enhancing the quantity or quality of food.

The public health costs of any reductions or delays—which would, at least for a period, make food more costly and reduce broader access to higher-quality food— would be disproportionately borne by the poorest and most vulnerable segments of society. In addition, the environmental gains flowing from a "reasonable" precaution should more than offset the environmental gains that would otherwise be obtained. Notably, the Universal Declaration on Human Rights, passed by the United Nations General Assembly in 1948, recognizes the importance of adequate food, clothing, and housing. Article 25(1) states that "everyone has the right to a standard of living adequate for the health and well-being of himself and of his family, including food. clothing, housing....".

The Universal Declaration on the Eradication of Hunger and Malnutrition is much more explicit. It specifies that "every man, woman and child has the inalienable right to be free from hunger and malnutrition." Thus, it could be argued that any efforts to block research, development, and, where appropriate, commercialization of GM crops would violate Article 10 of the Declaration on Human Rights, which states that "no one shall participate, by act or failure to act where required, in violating human rights and fundamental freedoms."

In summary, the rewards of GM crops greatly outweigh their risks. Although it would be a mistake to go full steam ahead on GM crops, it would be a bigger mistake to stop them in their tracks. The wisest policy would be to go as fast as possible while keeping a sharp lookout, and staying on the track to improvements in human and environmental well-being. It could be argued that a policy that would ban GM crops would violate the basic human rights of hundreds of millions who are not yet free from hunger and malnutrition.

7

Biotechnology vs Ethics and Politics

ETHICS TO POLITICS

Arguments were developed to show how our relationship with the natural world might be understood philosophically and ethically. What becomes clear is that it is extremely difficult to accurately describe the ethical relationships involved and the duties and obligations that these generate. That such ethical considerations are central to the commitment of many green activists is not in doubt; the point now at issue is whether political strategies and institutional arrangements can be derived directly from such reflections and commitments.

Could the belief in, for instance, the intrinsic value of all natural entities tell us anything about the ideal form of institutions? Is it possible to 'read off' political arrangements from reflections on natural processes? For example, is it plausible for political ecologists to claim that diversity in nature equates to toleration; stability to democracy; interdependence to equality; and longevity to tradition? Are these claims that features of the natural world equate to social and political forms and principles coherent?

There are a number of sceptical points that need to be made here which are crucial to the development of green politics. The first is as true for green politics as it is for environmental ethics: it is simply not possible to 'read off' ethical and political principles direct from scientific concepts such as diversity, symbiosis or complexity. Normative concepts cannot be drawn directly from descriptive ecological concepts. To argue that we can 'because they are natural' is to argue in a circle. As John Barry comments, 'non-human nature gives us no determinate prescriptions about how we ought to live'. However, there may be compelling practical political reasons to argue that nature does provide a guide to action. It is important to realise that reading off from nature may be 'meaningless in political-theoretical terms, but useful from an ideological point of view where persuasion is so important ... the symbolic potency acquired by doing so may make the price of vagueness worth paying'.

Again, it is far from obvious that ethical commitments to the non-human world, such as a belief in the moral worth of sentience or the intrinsic value of aspects of the non-human world, can tell us anything about whether we should commend democratic or authoritarian solutions to political questions. The link between ethical and political commitments is rarely clear-cut. In fact it is a fair criticism of environmental philosophy that it provides little in the way of *practical* guidance in the area of political change. This is not to say that ethical reasoning and ecological insights are a waste of time and play no part in political considerations.

All that is being stressed is that there is no *necessary* one-to-one connection between ethical and political commitments. What such ethical reasoning can provide us with, however, is a critical standpoint from which to assess political arrangements and decision-making processes. Such reflections can help us to make judgements about the environmental sensitivity of institutional designs; for instance, in contributing towards criticism of purely economic decision-making processes such as cost-benefit analysis (CBA) which cannot adequately represent such ethical commitments.

Considering these issues may help to articulate the point that is being made here. Such considerations do not simply stem from a particular view of values in nature, from, for example, the assertion of intrinsic value; but from a commitment to justice - a typicaliy anthropocentric concern. Studying ecological relations will not lead us to a precise definition of what our duties and obligations to present and future generations entail; that requires adherence to beliefs beyond such reflections. However, environmental considerations and values help us to deepen our understanding of humanity's relations with the non-human world and of our political and social institutions, but *without* offering us the basis for a complete blueprint for a future sustainable society.

The Limits to Growth Debate

If an ethical commitment to non-human nature provides a basic building-block of green political thought, the debate over the ecological limits of social and economic arrangements has provided another. In 1972, the Club of Rome, a group of prominent scientists, educators, economists, humanists, industrialists and national and international civil servants, published its report *The Limits to Growth*. The report was based on an investigation of the interconnected nature of five trends that the Club of Rome believed to be of global concern: 'accelerating industrialisation, rapid population growth, widespread malnutrition, depletion of nonrenewable resources, and a deteriorating environment'.

Using a computer-generated world model, a series of scenarios was developed by inputting different rates of change for each factor. Central to the report (and to much green analysis that followed) is the emphasis on the interrelatedness of different trends and the exponential, as opposed to linear, pattern of growth associated with these trends. A particular characteristic of such exponential growth is the suddenness with which it approaches fixed limits.

It is this characteristic that led to the pessimistic conclusion of the Club of Rome that the post-war rate of economic expansion and population growth could not be sustained without widespread poverty and famine, exhaustion of global natural resources and irreparable environmental damage. There is a decidedly neo-Malthusian feel to *The Limits to Growth* in line with the concerns for population growth expressed in books such as *The Population Bomb* by Paul Ehrlich, published in the same year.

1. If the present growth trends in world population, industrialisation, food production, and resource depletion continue unchanged, the limits to growth on this planet will be reached sometime within the next hundred years. The most probable result will be a rather sudden and uncontrollable decline in both population and industrial capacity.
2. It is possible to alter these growth trends and to establish a condition of ecological and economic stability that is sustainable far into the future. The state of global equilibrium could be designed so that the basic needs of each person on earth are satisfied and each person has an equal opportunity to realise his individual potential.
3. If the world's people decide to strive for the second outcome rather than the first, the sooner they begin working to attain it, the greater will be their chances of success.

Since its publication, the report has been widely criticised and a number of its predictions have failed to materialise. In retrospect, some of its conclusions and modelling proved overly simplistic and pessimistic. Julian Simon and Herman Kahn took the view in *The Resourceful Earth* (1984) that the proponents of the limits to growth thesis were guilty of gross exaggeration at best and scare-mongering at worst. Later authors such as the economist Wilfred Beckerman, in his ironically titled *Small is Stupid*, are equally scathing about such resource-depletion scare stories, pointing out that *The Limits to Growth* failed to allow for the various feedback mechanisms operative within an economy - if a good or resource becomes scarce its price rises which signals a search for new sources of that good or for suitable substitutes. Beckerman's critique in many ways exposes the overly pessimistic outlook of the Club of Rome's report, although it clearly requires there to be suitable substitutes: a claim that may not always be true of certain environmental goods critical to human survival.

Regardless of debates concerning specific resources, ultimately economic activity depends on the health of the environment; failure to recognise this could lead to devastating consequences. While we must recognise its rather overstated conclusions and simplistic computer modelling, the Club of Rome report nonetheless managed to focus attention on growing global tensions surrounding resource depletion. Its central theme - that infinite growth in a finite system is impossible - has become a 'foundation stone of Green political thinking'.

In the same year, the editorial board of *The Ecologist* magazine co-wrote *A Blueprint for Survival* and this was followed in 1973 by E.F. Schumacher's *Small is Beautiful*. These two seminal texts share with the Club of Rome a deep concern with the impact of human activity on the global environment. However, the solutions they suggest are significantly different. The Club of Rome argues for a concerted global response led by the major institutions of industrial societies; in contrast Schumacher and *The Ecologist* both emphasise the need to restructure society and economy on a 'human scale'. In their vision, a concerted move towards decentralisation is fundamental if society is to become more congenial and develop within ecological limits. These different visions lead to different trajectories within green politics, a fact commented on by Maarten Hajer:

> *Limits* argued for a further integration and hierarchisation in order to contain the problem (global problems required global solutions) while *Blueprint* argued for decentralisation, self-sufficiency, and self government; *Limits* accentuated the historical responsibility of social elites of business, science, and government, *Blueprint* came out in favour of self-government...whereas *Limits to Growth* typified very much the response to the increased importance of environmental matters from the world of business and government elites, *Blueprint* became a key reference for the radical environmental public of its time.

Whichever trajectory we follow, the limits to growth thesis remains central to the green standpoint. Indeed the thesis is taken further by greens who wish to emphasise the *moral* limits to growth, and mount a socio-economic critique of the inequalities and environmental insensitivity of liberal capitalist political economy. It is also important to note that there is no necessary connection between the two core principles of green political thought: the ethical basis of our obligations to the non-human world and acceptance of limits to growth. This very issue leads to tensions between some ecosocialists who have accepted that there must be ecological restrictions to the forces of production and greens who believe that they must also widen their ethical commitments beyond purely human well-being.

Bearing in mind the distinctions drawn above, we can begin to understand why there are tensions within green political thought between ecocentric and technocentric attitudes and between decentralised and highly coordinated institutional responses to environmental problems. It is to such differences and tensions that we now turn.

The Emergence of a Green Political Ideology

Any survey of green political texts from the last two or three decades will reveal that there is a wide variety of political institutions and strategies for change endorsed as the way forward to a sustainable future. From democratic to authoritarian regimes, centralised states to decentralised communities, planned to free-market economies, party politics to grassroots activism - all have found support at some time or another. Such endorsements have often been a response to particular contexts and situations: Robyn Eckersley, for instance, highlights three interconnected 'preoccupations' within green political thought which can be seen as a response to particular circumstances.

The first preoccupation, the crisis of participation, evolved primarily during the 1960s in the civil rights movement. Its central concern is with participation in decision-making processes and issues of resource distribution, leading to a recognition of democracy and social justice as important themes within the green movement. The second preoccupation, the crisis of survival, was originally inspired by the publication of *The Limits to Growth* and *A Blueprint for Survival* and 'marked the emergence of the *global* dimensions of environmental degradation and the *common fate* of humanity'.

The apparent urgency of the survivalist message led many writers to call for authoritarian solutions to the environmental crisis, thereby contradicting the democratic thrust of earlier commentators. The third and final preoccupation is the belief that environmental concerns are as much a crisis of culture and human character as a crisis of nature or the use of natural resources. If this is so, a broader ecological understanding of human needs, technology, and self-image becomes essential with a recognition that it is in our interests, understood in a comprehensive manner, to become less dependent on technological responses to environmental and social problems.

These preoccupations need to be recognised and each given due weight: but it is important to realise that they may be in tension with each other. For example, green concerns often stress the need for a holistic or organicistic approach to the solution of environmental problems, but these may in turn be markedly at odds with the concern for egalitarianism and the associated demands for social change which also form part of the agenda. If these tensions can be successfully addressed, green politics might then be linked with other emancipatory political projects that call for cultural renewal and the revitalisation of civil society. As Eckersley has argued:

> This new theoretical project is concerned to find ways of overcoming the destructive logic of capital accumulation, the acquisitive values of consumer society, and, more generally, all systems of domination (including class domination, patriarchy, imperialism, totalitarianism, and the domination of nature).

This is an extremely ambitious and radical remit for any political project and it would be an oversimplification to think that the values and attitudes it embodies have been developed purely from an analysis of our ethical relationship with the natural world or from the need to live within ecological limits. Equally, to view these developments as having occurred only within the last three decades would ignore a long history of attitudes to nature which have played a part in various streams of political thinking. As Andrew Vincent points out, the attitudes associated with ecology did not spring upon us in the 1970s with pure radical credentials.

Rather, they relate to a subtle and immensely potent conjunction of attitudes to nature which have been present in European thought since the late nineteenth century. Despite their widespread promotion by different and politically diverse groups throughout the twentieth century, it is the accidental conjunction of circumstances, individuals and events in the 1970s which has provided a dynamic refocus for the ecological vocabulary.

We need briefly to survey some of the traditions within Western political thought in order to understand how they have helped shape green political thinking. All the traditions have something to say about human-non-human relations and, in some cases, critical reassessment has occurred in the light of green critiques. One of the central issues we need to consider is whether green political thought has developed an entirely autonomous response to our circumstances or whether it still depends on the insights of other political streams.

We shall sketch historical associations between traditions and ecological thinking and highlight tensions to be found with current environmental concerns, specifically the ethical considerations of the non-human world and questions arising from the limits to growth debate. We shall also consider whether the central values of these traditions are compatible with green thinking. Ecological politics has developed from a critical relationship with many other streams of thought, and because of this it incorporates a number of diverse and contradictory tendencies.

Authoritarianism and Fascism

If scarcity is not dead, if it is in fact with us in a seemingly much more intense form than ever before in human history, how can we avoid reaching the conclusion that Leviathan is inevitable? Given current levels of population and technology, I do not believe that we can. Hobbes shows why a spaceship earth must have a captain. Otherwise, the collective selfishness and irresponsibility produced by the tragedy of the commons will destroy the spaceship, and any sacrifice of freedom by the crew members is clearly the lesser of two evils.

As this quotation shows, it is very easy to move from a perception of the seriousness of environmental problems towards a view in which, as William Ophuls argues, the choice is between 'Leviathan and oblivion'. In responding to the problems identified by *The Limits to Growth* and similar reports, writers such as Ophuls and Robert Heilbroner, while at a personal level committed to liberal and democratic values, envisaged an environmentally-benevolent Leviathan as the only political arrangement that could stop human societies from developing beyond environmental-carrying capacities. Restrictions on levels of production, consumption and population growth could not be achieved quickly enough through democratic processes, and individuals' rights and freedoms would have to be overridden in the short term in order to achieve long-term survival and lessen ecological damage.

Perhaps the most consistent and influential advocate of this position has been Garrett Hardin whose 'Tragedy of the Commons' thesis espouses strong, centralised leadership if self-interested individuals are not to despoil environmental resources. Along with his later, more controversial 'Lifeboat Ethics' - which seems to many to suggest that if the developed, Western nations are to survive then they should cut off aid links with the Third World and leave them in poverty - such authoritarian ideas have often found their way into green proposals concerning population control, resource distribution and immigration. It is essential for greens to realise that many of their concepts and much of their rhetoric can have authoritarian and imperialistic overtones.

Green theorists frequently appeal to concepts of the natural, using organic metaphors, elevating the spiritual over the rational, and justifying a holistic politics which elevates the community and the state over the individual. Such ecological doctrines have often been appealed to by authoritarian and fascistic movements: for instance, pantheistic ecological considerations were fundamental to the ideology of the Nazis. Although few greens would now espouse such political arrangements, they would do well to recognise that authoritarian solutions are often lurking in the background.

Conservatism

Many of the concerns of environmentalists appear to have their roots in central conceptions within traditional conservatism. Congruence can be found with ideas of tradition, continuity, stability, organic change, prudence, rejection of totalitarianism and appeals to community. Traditional conservative and ecological theorists often share an anti-capitalist stance and romantic visions of non-human nature. In a recent defence of the deep affinity between ecological theory and conservative philosophy, John Gray adopts a common criticism of neo-liberal market philosophy, stressing the similarities of the two streams of thought.

Both share a multi-generation perspective, give primacy to the common life, see danger in novelty and give a central place to the virtue of prudence. There is considerable scepticism about the possibility, inevitability or desirability of 'progress' and an emphasis on continuity and change as occurring within a developing tradition rather than in the light of a rational blueprint for society. The ideals of harmony and stability are central conceptions to both traditional conservatism and green political thought. As Michael Freeden remarks, 'conservative arguments cannot be completely disentangled or excluded from all green positions'.

Although there are indeed many areas of congruence, Gray's analysis can easily be construed as focusing selectively on particular areas of green thought and conservatism. Clearly the conservation and preservation streams of the environmental movement owe much to conservative and romantic visions of nature. It is often the case that in specific local environmental campaigns traditional conservatives are found in alliance with more radical, emancipatory greens. However, contemporary green analysis of social relations appears to part company with much conservative thought on various issues, although even here the case is not clear cut.

Where conservative thinking tends to emphasise order, tradition and community over what it sees as abstract criteria such as social justice and egalitarianism, contemporary green political analysis typically takes these later ideas as central and is critical of apparent conservative apologies for hierarchical social arrangement.

Here we find a potential paradox in green thinking, with calls for universal standards of justice and democracy coexisting with a desire to defend indigenous communities and their practices. But what if those practices, which may be environmentally sustainable, are themselves based on the dominance of a particular social group? It is with such questions that green political thought must come to terms and perhaps it would be more accurate to see conservatism and romanticism as modes of thought out of which a more comprehensive green critique is developing.

Liberalism

Environmental thought again shares a mixed relationship with the liberal tradition. Although influenced in many ways by early liberal thought, particularly ideas of rights, freedom and democracy, the most recent turn in liberal thought, namely neoliberalism and laissez-faire free-market economics, appears incompatible with the central tenets of green thinking. The influence of two early liberals, Jeremy Bentham and John Stuart Mill, has already been noted. Bentham's utilitarian views on public policy have found favour among animal rights theorists where sentience and the ability to suffer pain are central to considerations of the proper treatment of animals. Many of

Mill's ideas are similarly resurfacing in green texts. Herman Daly, for instance, has championed the steady-state economy which was first adumbrated in Mill's *Principles of Political Economy*. Recent work has attempted to demonstrate that liberalism and environmentalism are far from incompatible and liberal approaches to environmental problems have been used to argue for the extension of rights to both future generations and to other living things.

Appeals to nineteenth-century liberal ideals such as public space and deliberation, civic virtue and civil society, show that certain streams of liberal thought are central to green visions of future sustainable societies. However, it would be fair to say that where many connections can be made to these classic liberal values, the rise of neo-liberal or New Right thought in recent times has been viewed with horror by the majority of environmentalists.

Individualism, the pursuit of private gain, limited government and market freedom are contradicted by radical ecology commitments to the resolution of environmental problems as a collective good and to the intervention and restrictions on economic and personal fieedoms to deal with them. Liberal economy is seen to underpin the commitment to economic expansion and accumulation and to the identification of wealth and material advancement with progress and improvement.

Such material accumulation is seen by many green writers as spelling ruin and destruction for human and ecological communities, putting increasing pressure on social and environmental relations. Further, the neoliberal conception of well-being as a correlative of material acquisition is viewed as a complete misrepresentation of what is important to a human life. There would appear to be serious contradictions between neo-liberal market logic and environmental imperatives. Recently, though, there has been a response to such criticisms by advocates of 'free-market environmentalism'. For such neo-liberals, ecological problems such as Hardin's 'Tragedy of the Commons' are the result of a lack of well-defined property rights and price mechanisms; rather than reducing the impact of the market we need to extend its role.

This is very much a counter move by the New Right in order to deflect greens' anti-free market claims and to rework environmental problems in neo-liberal rhetoric. It is important, however, to note that the rejection of a free market approach to environmental problems does not necessarily mean that greens must be totally anti-market: the market and price incentives and disincentives can be used as part of environmental policy without being seen as the *whole* of policy.

In more general terms, the individualism which lies at the heart of liberalism, whether in its political expression (the insistence on rights and liberties) or in its economic expression (the insistence on markets), creates a problem for environmentalists who are attempting to develop new forms of community and political participation relevant to a sustainable society. Much green thinking has a greater affinity with various forms of communitarianism, both in its insistence on the essential relationship of an individual to a society (as opposed to the liberal conception of society as being nothing more than the creation of pre-existing bearers of rights) and in its insistence on responsibilities rather than merely on the assertion of rights.

Marxism and Socialism

It is very common to find the criticisms levelled at capitalism and liberal free-market ideology mirrored in discussions on socialism and communism. As Jonathan Porritt argues: 'Both [capitalism and socialism] are dedicated to industrial growth, to the expansion of the means of production, to the materialistic ethic as the best means of meeting people's needs, and to unimpeded technological development'. For writers such as Porritt, both are forms of the 'super-ideology' of industrialism, with the former communist states displaying a record of environmental protection as bad as (if not worse than) Western nations. Although socialists share green concerns for poverty relief and egalitarian redistribution of resources, increased wealth generation is often seen as essential in order to finance such policies.

Greens argue that socialist thought does not take into account the physical limits to wealth production, seemingly believing in the ability of science and technology to overcome scarcity. Further, it is argued that the environment is only considered in terms of its direct impact on the well-being of humans. Socialists have responded by accusing greens of having an idealised vision of the non-human world and providing 'in essence a defence of middle-class privilege and exploitative class and neo-colonialist relationships'.

However, between these rather polarised positions the continuing debate between greens and socialists has in the last few years resulted in the re-examination of their shared attitudes in light of the apparent hegemony of neo-liberal thinking. It is both simplistic and narrow either to take the former communist bloc as the basis of criticism of all Marxist and socialist ideas, or to view greens as only interested in the ecological effect of production.

There are at least four areas where fruitful connections have been made and where the green analysis has been deepened by socialist reflections - and vice versa. First, in the area of political economy, socialist thought exposes in some depth the destructive power of capitalist societies. Such an analysis has helped deepen and refocus the ecological critique of industrialism. Socialists have often adopted the sceptical view that greens have seemed willing to ameliorate the ecological impact of capitalism rather than call for a full-scale transformation of economic structures.

As André Gorz argues:

It is...time to end the pretence that ecology is, by itself, sufficient: *the ecological movement is not an end in itself, but a stage in the larger struggle.* It can throw up obstacles to capitalist development and force a number of changes. But when, after exhausting every means of coercion and deceit, capitalism begins to work its way out of the ecological impasse, it will assimilate ecological necessities as technical constraints, and adapt the conditions of exploitation to them.

The centrality of growth, the injustice of the existing distribution of wealth, and the short-term interest in profit maximisation of free-market ideology might be seen as common elements within socialist and green thought. However, for more mainstream socialists the generation of material wealth is not in itself a problem, rather the method of its production and distribution. Here we do find a tension between those socialists who are willing to accept the ideas of natural limits to economic expansion and an ecological dimension in political economy and those who seem overly optimistic and share the Promethean belief that the problems of poverty, injustice and inequality can be solved through the abolition of scarcity by technical means.

Marx largely viewed nature as something to be dominated and used for human purposes. Where he identified and denounced exploitation it was the exploitation of class by class through their different places within the relations of production (the pattern of ownership, property, law and control over the forces of production); the forces of production and their domination over nature he much admired. The task, as he saw it, was to harness the technological possibilities of exploiting nature in the service of a fundamental change in the social relations of production in which the surplus labour value of the worker no longer accrued solely to the benefit of the capitalist class.

For Marx, the question of the desirability of exploiting nature, or the limits to such exploitation, were not a matter of concern, and nature as such was not ascribed value independently of its instrumental value to human beings in allowing them to develop and progress. The ecological dimension cannot, therefore, be straightforwardly grafted onto orthodox Marxism, requiring as it does a reorientation of human relations with the natural world.

The second area of interest is the question of agents of change. Although often providing clear prescriptions for a future sustainable society, green political thought has often been rightly criticised for inadequately theorising strategies for change. On the other hand, orthodox Marxism has a ready-made explanation of both the causes of change and the agents of change. With its emphasis on the forces of production and its analysis of the social relations of production, it could confidently state that the main agent of change would inevitably be the working class.

But the Left is now in a position of turmoil on this issue, with the traditional Marxist emphasis on the working class as the vanguard of political revolution becoming increasingly unconvincing. Both green and socialist writers have turned their attention to political strategies that incorporate, for instance, new social movements, the unemployed, communes, and more traditional parliamentary routes.

The third area of critical debate centres on the role of institutions, particularly the state, both in processes of change and in any future society. Many ecosocialists tend to view the green commitment to a decentralised society as a romantic, utopian ideal that is doomed to failure, would result in further inequalities, and be unable to tackle large-scale environmental problems because of a lack of coordination. An enlightened state is often seen to be essential if a sustainable future is to be achieved.

If one is honest, however, about the objectives which an ecologically enlightened state would set itself, it is difficult to avoid concluding that the state, as the agent of the collective will, would have to take an active law-making and enforcing role in imposing a range of environmental and resource constraints.

It is worth noting however that there are streams of socialist thought based on decentralist values which do not put so much stress on statism. Writers such as William Morris, Erich Fromm and Ivan Illich, whose work has often been ignored within left-wing thought dominated by reformist and revolutionary statist political thinking, are now being reinterpreted and given their due in the light of the ecological imperative to restructure societies.

Finally, much has been written about the Marxist attitude to the non-human world. The early writings of Marx can be interpreted as showing an awareness of the interconnectedness and dialectical relationship of humanity with the rest of nature and the alienation of both under systems of capitalism. However, even with such a positive interpretation and the concept of nature as 'man's inorganic body', it is clear that Marx's dialectics are orientated specifically at the self-realisation of humanity through the domination and transformation of nature.

Socialists are often charged with being interested in environmental concerns only in so far as there is a direct effect on human well-being. In response, greens are often vilified for a tendency towards mysticism and romanticism. But do differences in orientation of this sort necessarily issue in differing policy implications? For the most part it might not seem so. However, it is easy to see how on specific issues, for example species extinction, there might be no relation between human well-being and the threatened species.

Socialists could then apparently offer no reason for protection. But this may not be the final position: socialist beliefs do not necessarily rule out consideration of non-human entities, even if in practice they have focused their attention primarily on human welfare. There are certainly a number of post-Marxists, such as Herbert Marcuse, who recognise the link between the domination of man by man and nature by man and look forward to the re-enchantment of the non-human world.

More recently, writers such as André Gorz, David Pepper, Ted Benton, Tim Hayward, Michael Jacobs and James O'Connor have done much to reconcile the apparent differences between socialist and green political thought. The ecological dimension to political economy has been generally accepted as a paramount concern and 'red-green' dialogue on political, social and economic arrangements is an area of extremely fruitful work.

The political arrangement supported by many radical greens is along anarchist lines, where the ideals of egalitarianism, non-hierarchy, local empowerment and democracy, self-reliance and diversity are often central concerns. This arises out of a deep mistrust of traditional forms and bearers of political authority, and a desire to challenge them and to re-assert that power and authority lies with people and their communities. The writings of late nineteenth- and early twentieth-century anarchists, such as Peter Kropotkin, can be seen as precursors to the more recent *A Blueprint for Survival*. Anarchist arguments for decentralisation of social and political institutions are taken to be absolutely necessary if a sustainable society is to emerge. This vision can be contrasted with some of the more authoritarian responses to *The Limits to Growth* report discussed earlier.

As with other traditions of thought, the ecoanarchist stream is not unified and can be subdivided into social ecology and ecocommunalism. The most influential social ecologist is undoubtedly Murray Bookchin whose analysis of environmental problems is based on a perceived connection between the exploitation of nature and the exploitation of human beings.

Contemporary society is seen as socially, economically and politically irrational and anti-ecological in that the competitive capitalist system is based on hierarchical relations of command and obedience within which nature is commodified. It is only when the domination of human over human is overcome that the domination of nature can be transcended. This will only occur in decentralised, autonomous and radically democratic communities.

Such an analysis is very different from that of the ecocommunalists who draw extensively on deep ecology. Here the humanity-nature relation is central, sometimes to the exclusion of relations between humans. Under the influence of deep ecological insights, bioregionalists, such as Kirkpatrick Sale, hold that communities must learn to live within the carrying capacity of their specific bioregion. In many ways there is a resonance within bioregionalism of certain 1970s survivalist views that self-sufficient, small-scale communities are the only feasible response to an imminent environmental catastrophe.

The deep ecological position is frequently criticised for simply 'reading off' ethical and political principles from ecological concepts such as diversity, symbiosis and complexity: on what grounds can we derive ethical or political principles direct from such scientific concepts? Arne Naess's conception of deep ecology, and particularly his commitment to 'biospherical egalitarianism', has been vociferously challenged by Bookchin who argues, at times polemically, that the de-centring of humans in such a political position has a tendency towards anti-humanism and misanthropy.

For Bookchin, too much emphasis is placed on the mystical qualities of the human-non-human relationship, on the development of an ecological self or consciousness, while the social, political and economic roots and realities of the environmental crisis are ignored. It is true that both social and deep ecology wish to confer the 'maximum political and economic autonomy on decentralised local communities', that their 'anarchism is grounded in, or otherwise draws its inspiration from, ecology' and that both positions provide a strong defence of 'the grass roots and extra-parliamentary activities of the Green movement', but the important differences between deep and social ecology must also be recognised.

The anarchist belief that the state must be either bypassed or abolished leads to much heated debate within green political thought. The state is viewed as the prime example of an oppressive agent and a defender of hierarchical social, economic and political power relations. Also it is seen to be too far removed and centralised to deal with local environmental and social problems.

However, for many critics, and particularly ecosocialists, the removal of the state and subsequent emergence of autonomous communities is viewed as overly idealistic and utopian and fails to take account of fundamentally important state functions. First, there is some concern over how the internal political structure of decentralised communities would be controlled.

For social ecologists, radical participatory democracy is fundamental. But, for bioregionalists it is living within the ecological-carrying capacity of a bioregion that is of prime importance rather than the political arrangements of a society. This is clearly implied by Sale:

> Bioregional diversity means exactly that. It does not mean that every region of the north-east, or North America, or the globe, will construct itself upon the values of democracy, equality, liberty, freedom, justice, and other such like *desiderata*. It means rather that truly autonomous bioregions will likely go their own separate ways and end up with some quite disparate political systems - some democracies, no doubt, some direct, some representative, some federative, but undoubtedly all kinds of aristocracies, theocracies, principalities, margravates, duchies and palatinates as well. And some with values, beliefs, standards and customs quite antithetical to those that the people in this room, for example, hold dearest.

Thus the stress placed on the value of diversity can be interpreted to mean a radical diversity of institutional forms. Hence, some of the more politically-sensitive questions concerning authoritarian solutions to population, immigration, punishment and the like return to the fore. There is a further issue surrounding the value of diversity which is so central to the rhetoric of anarchists - could a small-scale society support physical and cultural diversity and difference internally? Do not small, self-contained communities often lead to conformity and ostracise difference? In this respect the state may be viewed as the guarantor of democracy and the defender of political freedoms and difference.

Second, there is a question of distribution of resources: is it just that some communities or bioregions would be environmentally more abundant than others? Is there not some obligation for those who live in resource-rich locations to redistribute to those less fortunate? A similar issue is raised in terms of transboundary and large-scale environmental impacts.

A small-scale community is likely to respond to local environmental problems more sensitively than a more centralised authority, but without some form of over-arching authority, how are cumulative, supra-community problems such as acid rain or climate change to be faced? How are claims between communities over the impact of transboundary pollution to be settled? Finally, the question of strategy is underdeveloped. How are we to get to this utopia of decentralised communities from the present state structure and the increasing globalisation of capital and culture?

Many radical greens are fundamentally at odds with the idea of achieving this outcome through parliamentary means. Such processes necessitate concessions and leave too much power with too few individuals. If the means must be consistent with the ends then it is no surprise that potentially corrupting processes are vilified and in general lifestyle changes and the development of alternative communities and practices are endorsed. Although the wider political significance of such extra-parliamentary activities may at times be limited, the influence of anarchist thought, with its belief in local political action and its challenging of existing institutions and the interests they represent, is evident in much green political thinking.

Feminism

'Ecological feminism' is an umbrella term which captures a variety of multicultural perspectives on the nature of the connections *within* social systems of domination between those humans in subordinate positions, particularly women, and the domination of non-human nature.... Ecofeminist analyses of the twin dominations of women and nature include considerations of the domination of people of colour, children, and the underclass.

Ecofeminist analysis shares the broad concern of the feminist movement as regards the elimination of gender bias

- or more widely, all forms of bias towards sub-ordinated groups - and the development of non-gendered or non-interest dominated political, social and economic practices and institutions. Values and characteristics usually conceived of as feminine, such as humility, care and nurture, are seen to be seriously undervalued in patriarchal societies with a detrimental effect on how we conceive of our relationships with non-human nature.

In addition, ecofeminists argue that there are connections between the domination and oppression of women and of nature. This leads not only to a variety of ecofeminist positions that have strong political connections to existing political traditions, for example liberal, Marxist and socialist feminisms, but also variations based on the interpretation of the link between the domination of women and nature.

The various accounts of the source of the links between the subordination of women and the domination of the natural world seem to fit into two main categories. First, there is an argument that dominant philosophies have traditionally operated on the basis of various dualisms in which women and nature are both allocated an inferior position with respect to men and their projects. It is argued that this differentiation has had real historical consequences across all forms of human knowledge and practice, be it in art, science, education, ethics or politics. These consequences have led to attitudes justifying the exploitation of women and the natural world and can be seen as an expression of what is often termed strong anthropocentrism. Second, there is an account of domination which stresses the physiological and psychological *difference* between men and women. Here women are taken to be biologically closer to the natural order by virtue, for example, of their capacity for childbirth and nurture.

It is these conceptual connections that in many cases cause tensions and disagreements both within ecofeminist analysis and within the wider feminist movement. The second stream of ecofeminism highlighted often promotes the idea of a biologically- and psychologically-determined connection between women and nature. Women must be in the vanguard of environmental change simply because they are closer to and understand the natural world in a way that men can never achieve.

Such biological and psychological determinism, however, emphasises a fundamental difference between the sexes which for many critics could be used as the basis for further domination and servitude. Clearly there are controversies within ecofeminist thought with other commentators simply wishing to argue that a more environmentally-sensitive society needs to articulate those principles, values and characteristics traditionally associated with the feminine, but potentially accessible to all humanity. This latter position is invaluable in highlighting the power relations that lead to the marginalisation and exclusion of women, indigenous peoples and others, and the possibilities for radical institutional change called for by many within the green movement.

A new ideology

Should green political thought be understood as an ideology in its own right? It is common to find writers claiming that green thinking represents an overcoming of Right-Left distinctions and, on occasions, that it represents a complete break from traditional political thought; for others it is a new ideology that can take its place alongside the more established ideologies such as conservatism, socialism, liberalism or anarchism. It would seem however that the answer to our question is problematised by the fact that there is wide disagreement as to the meaning of the term 'ideology': it is an essentially contested concept within political theory.

Michael Freeden analyses ideologies in terms of the core, adjacent and peripheral concepts of which they are composed. Given such an understanding, one should expect an overlap between different ideologies in that they may hold many concepts in common. However, the place assigned to these concepts within the whole, and hence their relative significance and meaning, will vary from ideology to ideology.

In analysing green ideology, then, one should not expect that all its values and principles will be completely different from other ideologies; rather, one should expect to find some features in common, but contained within a whole whose overall thrust is substantially different. In this chapter, we have argued that at the core of green political thinking is the belief that our ethical relations with non-human nature and the finite character of resources need to be central in political reflections; beyond these two core ideas, green political thought must rely on insights from other traditions. So, are these core themes enough sufficiently to demarcate green political thinking from other ideologies and doctrines?

Of the green political theorists who have argued that it is possible to demarcate a green ideological position, one of the most significant and influential is Andrew Dobson. He argues that 'ecologism' represents a new ideology in its own right.

Ecologism provides a critique of contemporary political, economic and social arrangements and offers a prescription of how the world ought to be organised that differs fundamentally from the insights of other modern political ideologies. Dobson's position rests on an important distinction between 'ecologism' and 'environmentalism':

> *environmentalism* argues for a managerial approach to environmental problems, secure in the belief that they can be solved without fundamental changes in present values or patterns of production and consumption and *ecologism* holds that a sustainable and fulfilling existence presupposes radical changes in our relationship with the natural world, and in our mode of social and political life.

The central claim in his book *Green Political Thought* is that environmentalism can be easily assimilated into more traditional ideologies (as we have witnessed above). This is simply not the case with ecologism. Briefly, ecologism argues for radically different political, economic and social arrangements based on a reorientation on the part of individuals and society towards non-human nature. The ecological society is post-industrial and small-scale. Politically it favours local face-to-face participatory democracy. The economic vision is one of self-reliance and labour-intensive processes based on low-impact technology. Although there are other conceptions of post-industrial utopias, the vision of ecologism is significantly different:

> Ecologism envisages a post-industrial future that is quite distinct from that with which we are most generally acquainted. While most post-industrial futures revolve around high-growth, high-technology, expanding services, greater leisure, and satisfaction conceived in material terms, ecologism's post-industrial society questions growth and technology, and suggests that the Good Life will involve more work and fewer material objects.

It is environmentalism (using Dobson's terms) that is having the greatest impact on present policies, practices and ideas. The vision of ecologism may simply be utopian. Nevertheless, Dobson argues that it provides the indispensable fundamentalist well of inspiration from which green activists, even the most reformist and respectable, need continually to draw. Green reformers need a radically different picture of post-industrial society, they need deep-ecological visionaries, they need the phantom studies of the sustainable society, and they need, paradoxically, occasionally to be brought down to earth and to be reminded about limits to growth. Dark-green politics remind reformists of where they want to go even if they don't really think they can get there. The ideology of ecologism is the conscience of green politics.

Dobson makes an important point here about the role of utopian visions in political movements. The question remains whether he has isolated the single green ideological vision. Alternatively he may be over-emphasising a particular stream *within* a wide range of possible coherent green political positions that can be understood as a reaction against the domination of a strong anthropocentric or technocentric attitude, where nature is simply seen as something to be mastered and controlled.

Might it not be better to understand green political thought as a critical perspective which has forced existing political traditions to undergo an internal analysis and re-think of their fundamental premises and concerns, as well as providing a critical space for convergence and debate between traditions? Or is this simply 'environmentalism'; a watering-down of radical green insights? Rather than attempting to define the definitive ideological form for green politics, we may be wise to be cautious and to reflect on Martin Ryle's warning that 'the mere invocation of "ecology", crucial as it is, does not in itself determine in a positive sense the future development of social and political reality'.

Tensions and mutual distrust, both theoretical and practical, abound within green political thought, for instance over the role of the state and the limits of decentralisation, or the identification of agents and processes of change. Political commitments such as justice, democracy and liberty cannot be developed from purely ecological considerations, although ecological ideas can have implications for their preferred form.

It is perhaps easier in many ways to draw out what contemporary green political thinking rejects, such as authoritarian solutions and free-market capitalism, than provide an unambiguous account of a unitary green ideological position. Contemporary green ideas can perhaps best be understood as developing from a fluid, critical reflection on traditional areas of political thought, particularly areas of socialist, anarchist and feminist thinking, with ecological concerns as a necessary, but not complete, part of that reflection.

Ecological Modernisation

Whether or not we are able to discern a definitive form of green political ideology, its privileged position within the writings of green theorists is certainly not mirrored in more mainstream discussions of environmental politics. It is the ideas and practices of sustainable development that dominate proceedings. It is indisputable that the concept of sustainable development grew out of green political thought. However, to what extent does green political ideology remain 'the indispensable fundamentalist well of inspiration' in current debates and practices?

As we saw in the Introduction, much rests on how the concept of sustainable development is to be interpreted and employed by those interested in environmental problems. Agreement on the importance of the concept of sustainable development can be found across the ideological spectrum. Groups as diverse as neo-liberal free-marketeers and radical bioregionalists agree that it is significant, but disagree about its implications. Consensus on the importance of a concept such as sustainable development can disguise the extent of the divergence in the ways in which it is understood by different people and interests. Different groups share the concept, but have different conceptions of sustainable development: sometimes it would seem that all there is in common between them is the very phrase 'sustainable development' itself. It is an essentially contested concept and institutional arrangements, policies and strategies for change do not simply follow from its invocation.

But although the concept of sustainable development is contested, contemporary debates are nonetheless dominated by a particular interpretation: ecological modernisation. Environmental politics raises the question of the compatibility of economic growth and environmental protection; sustainable development is the conceptual framework within which these questions are typically explored. Ecological modernisation adopts a distinctive stance on the question of economy-environment integration, challenging 'the fundamental assumption of conventional wisdom, namely that there is a zero-sum trade-off between economic prosperity and environmental concern'. Ecological modernisation, it is claimed, offers a 'win-win' scenario whereby economic growth and environmental protection can be reconciled. As Hajer states:

Ecological modernisation can be defined as the discourse that recognises the structural character of the environmental problematique but none the less assumes that existing political, economic, and social institutions can internalise the care for the environment.

The Brundtland Report is a seminal text in the development of ecological modernisation in that it emphasises the mutual reinforcing of economic growth, social development and environmental protection. As with *The Limits to Growth* report, environmental degradation is linked with patterns of economic development. However, Brundtland concluded that continued economic growth is essential for environmental protection. This interpretation of sustainable development has been widely endorsed at all levels. It is easy to understand why such a definition of sustainable development could be supported by many different parties (including governments, businesses, reform-minded environmentalists and scientists) in that it apparently offers the panacea of combining economic growth and environmental protection.

A theoretical and practical consensus is emerging around the discourse of ecological modernisation; this challenges the more radical green idea that fundamental reorganisation of the major institutions of modern society is a necessary precondition of long-term sustainable development. Ecological modernisation leaves essentially unchallenged the system of industrial production, the capitalist economy and the centralised state. The widespread political support for the conception of sustainable development as ecological modernisation is forcing many radical greens to disown the concept they originally coined. They do not want to be drawn into what they see as a false or misleading consensus which might serve to legitimise the existing institutional order and thereby blunt the edge of radical critique.

A number of commentators have attempted to characterise the core ideas within ecological modernisation theory. What emerges is that it is optimistic about the ability of policy makers, scientific experts and businesses to integrate ecological concerns into everyday practices. In the policy-making realm, the adoption of techniques and principles such as integrated pollution control (IPC), the precautionary principle and economic instruments are seen as ways of refocusing current economic development. The environment is no longer seen as a free good; markets are made sensitive to the cost of environmental destruction and pollution; and anticipation and prevention is preferred to simply reacting to environmental problems as they arise.

Features of Ecological Modernisation

1. New policy-making principles, e.g. from react and cure to anticipate and prevent; integrated pollution abatement; integrating environmental concerns into all ministries; techniques allowing firms to integrate environment into cost-risk calculations, such as polluter pays, cost-benefit analysis, risk analysis, precautionary principle, tradeable pollution rights, pollution charges and taxes.
2. A new role for science in policy making, especially ecological systems science. Experts take central control.
3. Environmental protection seen as a source of growth since low- and non-wasteful anticipatory technologies generate profit.
4. Nature reconceptualised as a public rather than a 'free' good, so efficiency involves internalising environmental costs.
5. Burden of proof reassigned to the suspected polluter, not to the damaged party.
6. Policy-making process opened up to include new participation and partnerships, e.g. between business and NGOs. Voluntary agreements rather than command-and-control regulation are encouraged.

Ecological modernisation is an attempt to increase the environmental efficiency of the economy both at the macro-economic level (by signalling a shift away from energy- and resource-intensive industries) and at the micro level (by moving from 'end-of-pipe' to clean technologies).

> Advocates of ecological modernisation argue that ... changes, at both macro-economic and micro-economic levels, have the potential to make significant improvements in the environmental performance of industrial economies. Ecological modernisation is thus presented as a means by which capitalism can accommodate the environmental challenge. Rather than environmental protection being a threat to capitalism, it is seen as a spur to a new phase of capitalist development.

It becomes clear why ecological modernisation has been so widely embraced. For governments it offers an alternative, integrative policy approach which compares favourably with earlier ad hoc, remedial attempts to deal with environmental problems and it allows them to 'square the circle' between previously antagonistic environmental and business interests. Environmental problems are no longer seen as in fundamental opposition to capitalism, and this clearly also allows the business world to embrace the idea: from being villains of the piece, businesses have a new and virtuous role as important actors in the development of an ecologically-modern society. Ecological modernisation offers businesses the possibility of cost savings in production, clean technology and new market opportunities in pollution-control equipment and other 'green' products. And for reform-minded environmental organisations there is the promise of improved environmental conditions, the environment is taken seriously by decision makers in the public and private sectors, and increased opportunities for access to the corridors of power emerge.

Ecological modernisation is more than an idea or theoretical construct. Its advocates argue that a number of advanced capitalist nations are already developing along these lines and that we are witnessing a 'decoupling' of economic growth from ecological damage. Over the past decade, Germany, the Netherlands, Japan and the Scandinavian block appear to have achieved above OECD average improvements across a range of industry-related national environmental indicators, including water quality and air pollution emissions. In these countries there is now evidence of a decoupling of

GNP growth from the growth of environmentally harmful effects, including increased economic output with decreased energy and materials consumption per unit of GNP.

This adds up to a challenge to the fundamental green critique of large-scale industrialised economies. The claim is that we can move beyond opposition and confrontation towards cooperation and mutual coexistence between the interests of capital and the interests of the environment. Ecological modernisation aims to 'green' capitalism. But, on closer inspection, does it fulfil its promise?

Challenging Ecological Modernisation

It will not come as a surprise that the concept of ecological modernisation (and the practical claims made on its behalf) have been the object of much analysis and debate within green politics. Its faith in environmental solutions rooted in continued (although re-focused) capitalist, industrialised, scientific and technological development immediately raises the suspicions of those sympathetic to a more fundamental green outlook. A number of distinct criticisms have emerged that question the truth of ecological modernisation's decoupling thesis, challenge its perceived neglect of issues of social justice, dispute its conceptualisation of nature, doubt its faith in technological and scientific progress and take issue with the role played by the green movement itself.

The decoupling thesis has been challenged, first, because many apparent examples of decoupling have only been achieved through a displacement of high energy-consuming and polluting industries to less-industrialised countries: quite simply, environmental damage has been exported. Furthermore, increased efficiency of resource use makes no practical difference if consumption of goods increases and the economy continues to grow.

Even in the few countries where industrial sectors have achieved absolute decoupling, the improvements have been small and environmental impact remains at unsustainable levels. It is also worth noting that decoupling (where it has happened) has not occurred in the systematic manner that ecological modernisation theory might suggest. It has largely been an unexpected and un anticipated consequence of contemporary changes in the global economy. The question thus still remains: can positive government intervention achieve similar outcomes over the long term?

Second, the theory of ecological modernisation is charged with a failure to address pressing issues of social justice, because environmental preservation is no longer regarded as fundamentally contradicting capitalist development. On this view, ecological modernisation ignores social contradictions. It ignores questions of justice on two fronts: both within industrialised nations and between highly industrialised and Third World nations.

It is argued that ecological modernisation theory is silent on questions of the distribution of environmental goods and bads, focusing primarily on the overall domestic environmental impact of national economies in the industrialised part of the world. Environmental justice campaigners within industrialised nations highlight the unequal distribution of environmental risks; for example, the siting of toxic waste-disposal facilities close to lower socioeconomic neighbourhoods.

On a more global level, we have already seen that the ecological modernisation of national economies may only be achievable through the export of polluting industries and the degradation of environmental resources in less-industrialised nations. For writers such as Martin Khor, we must not lose sight of the role that capitalism plays in the continuation of poverty and environmental degradation:

> This is the ultimate environmental and social tragedy of our age: the scientific knowledge that could be properly used to provide for every human being's physical needs is being applied instead through industrial technology to take away resources from the Third World largely for the production of superfluous goods. Meanwhile, the majority of Third World peoples sink deeper into the margins of survival.

Advocates of sustainable development have been criticised for promoting development patterns that are culturally and ecologically insensitive to the needs and values of less-industrialised nations. However, at least the concept of sustainable development has 'global reach' - its pretensions towards global responsibilities and interconnectedness offer opportunities for alternative non-technocratic and non-patriarchal conceptions of development to emerge.

Even the much maligned (by radical greens) Brundtland Report *Our Common Future* stresses the fulfilment of needs in its much-quoted definition of sustainable development: 'development that meets the needs of the present without compromising the ability of future generations to meet their own needs'. The very next sentence in the report emphasises that it is 'the concept of "needs", in particular the essential needs of the world's poor, to which overriding priority should be given'. But within the discourse of ecological modernisation questions of Third World development appear to be neglected: ecological modernisation is a discourse of (and for) highly industrialised nations.

Third, ecological modernisation fails to engage with the moral critique offered by greens. Admittedly nature is integrated in decision-making processes, particularly economic policy, and is no longer viewed as a free good. However, there remains a highly instrumental view of nature because the discourse is that of ecoefficiency in which a reorientation towards the natural world is promoted for reasons of efficiency and not ethics. This is manifest in the promotion of policy techniques such as extended cost-benefit analysis. Environmental impacts are internalised, but only in economic terms. Although there may be pragmatic arguments for this internalisation, greens typically argue that it misrepresents the different types of value that we associate with the non-human world. The nature of aesthetic and ethical values is misrepresented in economic terms. Further, such valuation appears to reinforce a technocentric worldview.

Fourth, the trajectory and characterisation of modernisation as being (or as leading to) ecological modernisation is questioned within sociological theory. Ulrich Beck's formulation of the *Risk Society* contends that with the emergence of potentially devastating technological advances, such as nuclear power and biotechnology, the attention of government has shifted from the distribution of wealth to the distribution of risks. The unintended and unexpected side effects of industrialisation, such as nuclear contamination, climate change and the impacts of genetic modification, potentially affect all - no one can escape. In the risk society everyone is a potential victim of ecological devastation. And it is the institutions of industrial society that are implicated in environmental damage: there is a loss of legitimacy and trust in politicians, scientists, businesses, etc., and widespread anxiety and disenchantment with scientific and technological progress. The very actors that are seen as fundamental to the establishment of ecological modernisation no longer enjoy public support.

Finally, there is a widespread suspicion among more radical greens that ecological modernisation is simply a strategy of political accommodation: 'a rhetorical ploy that tries to reconcile the irreconcilable (environment and development) only to take the wind out of the sails of "real" environmentalists'. Reform-minded environmentalists are enthusiastic in the greening of capitalism, but capitalism it remains. The danger is that the radical critique of the environmental movement is blunted (if not neutralised) and the very structures and institutions responsible for continuing ecological decline are legitimised by apparent green approval.

Ecological Democratisation

Ecological modernisation is one possible conception of environmentalism. It is a managerial approach that does not fundamentally challenge social, economic and political practices. Admittedly production and consumption are to be made more ecologically sensitive; but overall, the debate is technical rather than ethical, promoting technical responses to environmental problems. But if, as we have suggested, 'ecologism' is a merely utopian vision, are ecological modernisation and the darker risk society the only two conceivable and practicable possibilities?

An alternative trajectory appears to be emerging within contemporary green political theory, one that might be termed 'ecological democratisation'.

Although ecologism may provide succour for the imagination, its rejectionist stance towards 'modernisation' is typically found to be unappealing. However, the technocratic character of ecological modernisation is also deeply problematic, especially with the spectre of risk society looming large. And it is this weakness in ecological modernisation - the loss of legitimacy of the dominant institutions of industrial society in the eyes of the public - that offers a glimpse of an alternative to ecological modernisation. It is only through the democratisation of technological, economic and political decision making that the legitimacy of (and trust in) institutions will be rebuilt. A theory of ecological democratisation begins to emerge.

If we are to move towards a more sustainable and equitable future, social, political and economic institutions will need to adapt to new ways of doing things. However, contemporary institutions at all levels (global to local) lack legitimacy in that they are implicated in the growing disparity of wealth within and between societies, increased environmental degradation and the inability to act within the confines of the global capitalist system. Their practices and interpretations of sustainable development are seen to favour the interests of particular politically-influential groups within society and are relatively unconcerned with the experiences and needs of the disenfranchised. Unsurprisingly the motives of the state and other institutions are distrusted, leading to feelings of political alienation, cynicism, and general apathy towards political institutions.

It would seem, then, that the crisis to which contemporary green politics must respond can best be understood as a crisis of representation, or more accurately, a crisis of misrepresentation. Knowledge of, and ethical commitments to, non-human nature are frequently misrepresented in technical, economic and political decision-making processes by 'experts' and by our political representatives. How then, in such circumstances, can institutions be trusted to develop sustainable policies and practices? One answer is that meeting the demands of sustainable development requires radical political restructuring to reflect the experiences and needs of all sectors of society.

Opportunities need to be available to challenge the environmental and social implications of current practices. Decision-making processes need to be designed to ensure that the voices of the disenfranchised are heard and different interpretations of sustainable development and our relations with the natural world explored. Is it possible to design institutions to guard against the manipulation of the political agenda by powerful elites?

Certain key themes begin to emerge from our reflections on the nature of green politics and the connection between sustainable development and democracy. Contemporary capitalism primarily requires governments to protect private capital and to ensure continued economic growth. Under such conditions, environmental considerations (where they conflict with capital accumulation) are unlikely to be given priority. The demands of sustainable development appear to be at odds with the logic of capitalism. This may be a further reason for greens to adopt a more critical stance towards ecological modernisation as currently understood. Sustainable development points towards the priority of democracy over capital accumulation. It requires the democratic control of development patterns. Greens may have good reason to be democrats.

However, theorists such as Robert Goodin are more sceptical about the necessity of democracy within green political thought and practice. In his view, greens should be concerned primarily with achieving sustainable ends rather than with the justice of the means: 'To advocate democracy is to advocate procedures, to advocate environment is to advocate substantive outcomes: what guarantee can we have that the former procedures will yield the latter sort of outcomes?'. But he has overlooked precisely what we have stressed in this chapter, which is that environmental concern can lead to the advocacy of a number of substantive outcomes.

Greens subscribe to a range of values and practices and do not universally agree on a single blueprint for a sustainable society. If this is the case, the manner in which we debate which ends to pursue clearly becomes extremely important. Political spaces need to be opened up within which alternative conceptions of the future can be suggested and challenged. Thus questions concerning ends are (it turns out) inseparable from questions of means. Sustainable development is as much about democracy as it is about limits to growth and our ethical relationship with the non-human world.

The extension of different forms of democratic participation offers a strong alternative to ecological modernisation, an alternative which connects ecological thinking with other streams of emancipatory thought, such as critical theory and feminism. Making these connections is without doubt politically and theoretically important. As a result of the interchanges between greens and contemporary democratic theorists, alternative conceptions of democracy have emerged that are likely to be more sensitive to environmental issues.

Of particular interest is deliberative or discursive democratic theory which directly responds to the problems of representation and authority raised earlier. The core idea behind deliberative theories is that decisions should only be regarded as legitimate if they derive from a process of argument and deliberation in which all citizens have an equal right to be heard, in which arguments are won and decisions made only through the force of the better argument, and in which other forms of power and political influence derived from wealth or patronage have no place. Joshua Cohen provides a useful characterisation of deliberative politics:

> The notion of a deliberative democracy is rooted in the intuitive ideal of a democratic association in which the

justification of the terms and conditions of association proceeds through public argument and reasoning among equal citizens. Citizens in such an order share a commitment to the resolution of problems of collective choice through public reasoning, and regard their basic institutions as legitimate in so far as they establish the framework for free public deliberation.

Hence deliberative institutional designs are participatory in that the values and needs of all groups in society form the basis of political deliberation and the legitimacy of actions and institutions is rooted in reasoned agreement. Only when citizens are able to see that their political and ethical commitments to environmental issues are genuinely taken into account will they begin to regain trust and interest in political debate and action.

Greens committed to such a form of ecological democratisation emphasise a number of features of democratic deliberation. First, deliberative institutions are likely to be more 'ecologically rational' in that they have the ability to respond to the high levels of complexity, uncertainty and collective action problems associated with many contemporary environmental problems.

Deliberative institutions are seen as promoting improved information flows by actively engaging numerous voices, including those individuals and groups with direct experience of the effects of environmental change who are too often marginalised from political decision-making processes. When faced with high levels of uncertainty and risk, deliberative institutions promise an ingenious mechanism through which the application of scientific and technological knowledge and expertise might be democratically regulated - an institutional setting within which the barriers between 'expert' and 'lay' knowledge can be challenged and reformulated.

Second, democratic deliberation promotes the articulation of preferences and judgements that are 'public-spirited' in nature. In public discussion it is difficult to defend the typically narrow self-interested preferences of those who wish to despoil and degrade the environment. Democratic deliberation offers conditions within which conflict between different values, such as environmental protection and social justice, can be understood and explored and under which more environmentally-sensitive attitudes, practices and development paths can be promoted.

Third, democratic deliberation rests on an active notion of citizenship that might be the basis for political and cultural change and the development of a practice of ecological stewardship. Democratic deliberation offers conditions under which citizens will encounter and reflect upon ecological knowledge and values and are more likely to internalise these in their judgements and practices.

Such internalisation and social learning can profoundly reshape the boundaries of existing political citizenship beyond the nation state, generating 'additional and occasionally alternative transnational allegiances ranging from the bio-regional through to the global, as well as to other species and the survival of ecosystems'. Deliberative processes provide a conducive arena in which to expose citizens to alternative ways of conceptualising relations between human and non-human worlds.

Deliberative democracy does not necessarily require complete decentralisation, although more locally-based decision-making procedures would clearly allow a greater degree of participation. Deliberative democracy is not the same as direct democracy. The former does not necessarily require total participation of all citizens, but rather a reasoned decision-making process to which all voices have access. As has already been noted, larger-scale institutions are also necessary to tackle transboundary and global environmental problems. The contemporary world is characterised by the dispersed nature of information and knowledge and the plurality of values and commitments. Deliberative institutions would be designed to respond to this.

Deliberative democracy can also be seen as 'ecological' in so far as it is characterised by the notion of appropriate scale. The idea of deliberation cuts across existing political and economic boundaries, because, as Dryzek argues, 'the size and scope of institutions should match the size and scope of problems'.

Green politics responds to the lack of attention given on the one hand to the ethical standing of non-human nature and on the other to the limits to growth implicit in the finite nature of the planet. As we have seen, recognition of these concerns is insufficient in itself to generate a *comprehensive* political doctrine in its own right.

Greens necessarily need to engage in debates about the nature of democracy, justice, equality, etc. The outcome of these debates cannot be pre-determined by analysing the two core insights alone. Whether we believe that a distinct green ideological position has emerged or not, green politics provides powerful critical ground from which to challenge existing and potential ideas and practices. Nowhere is this critical engagement more necessary than with the analysis of the concept of ecological modernisation. Although greens are concerned about the way nature is conceptualised, much of their critique rests on commitments to social justice and effective democratic institutions.

However, the green analysis of ecological modernisation remains somewhat ambiguous. Ecological modernisation could be seen as an idea and practice that undermines more demanding visions of the green society and overwhelms the more fundamental green critque. Alternatively, as Hajer has asked, 'one may also wonder whether ecological modernisation does not in fact have a much more profound meaning and could be seen as the first step on a bridge that leads towards a new sort of sustainable modern society'.

Case Study: Principles and Policies of the Green Political Programme

The aim of this case study is to analyse the nature of the green political programme. What sort of policies do greens promote? On what principles are these policies based? The policy platform of the Green Party of England and Wales is used as the basis of this analysis.

Statement of Core Principles

Life on Earth is under immense pressure. The environment around us is threatened with massive destruction. Conventional politics has failed us because its values are fundamentally flawed. The Green Party isn't just another political party. Green politics is a new and radical kind of politics guided by these core principles:

1. Humankind depends on the diversity of the natural world for its existence. We do not believe that other species are expendable.
2. The Earth's physical resources are finite. We threaten our future if we try to live beyond those means, so we must build a sustainable society that guarantees our long-term future.
3. Every person should be entitled to basic material security as of right.
4. Our actions should take account of the well-being of other nations and future generations. We should not pursue our well-being to the detriment of theirs.
5. A healthy society is based on voluntary co-operation between empowered individuals in a democratic society, free from discrimination whether based on race, colour, sex, religion, national origin, social origin or any other prejudice.
6. We emphasise democratic participation and accountability by ensuring that decisions are taken at the closest practical level to those affected by them.
7. We look for non-violent solutions to conflict situations, which take into account the interests of minorities and future generations in order to achieve lasting settlements.
8. The success of a society cannot be measured by narrow economic indicators, but should take account of factors affecting the quality of life for all people: personal freedom, social equity, health, happiness and human fulfilment.

9. Electoral politics is not the only way to achieve change in society, and we will use a variety of methods to help to affect change, providing those methods do not conflict with our other core principles.
10. The Green Party puts changes in both values and lifestyles at the heart of the radical green agenda.

The first two core principles of the Green Party in the Statement above will be familiar by now: the recognition of

1. The values associated with the non-human world; and
2. The physical limits to growth. The relentless pursuit of material affluence and economic growth is seen as the root of both the unsustainable use of environmental resources and the failure to recognise the interdependence between humans and their environment. Changes in patterns of *consumption* and our understanding of well-being lie at the heart of the green political programme.

Traditional political platforms are concerned with disagreements over the *distribution* of resources. Distributional justice is key to the green programme, but policies also call into question what is being produced, how and why. Our current lifestyles and attitudes and the structure of our political, social and economic institutions are implicated in the environmentally- and socially-destructive aspects of modern capitalist economies.

Many aspects of the green programme are directly related to the first two core principles and primarily promote the conservation of resources. Capitalist economies tend to be linear in the sense that resources are depleted, products produced and consumed and finally disposed of into the environment as waste, frequently causing pollution. Greens argue that this linear process, which as economies grow becomes more intense, needs to become circular and reduced in size. Products should be more durable and when their initial use is over, they should not be disposed of, but rather repaired or recycled - utilised as resources for other production processes. There is a hierarchy for resource consumption and waste management which would be central to a green economy. continued The waste-management hierarchy

1. Reduction of overall consumption
2. Selective consumption - maximum use of secondary materials, durable, reparable and recyclable products
3. Waste minimisation
4. Re-use
5. Recycling (including composting)
6. Disposal to landfill only as last resort (long-term policy to phase out completely)

It is interesting that recycling, which is perhaps one of the most popularised (shallow) 'green' proposals, is not the primary focus of waste management. Recycling can often be seen as an attempted technological fix to make capitalist economies less environmentally damaging. A green economy considers recycling only as an option after reduction of consumption, re-use and the like.

The rise in 'green consumerism' is regarded in a similar way to recycling - at best only a partial improvement on present consumption patterns and at worst an attempt to make capitalism appear more ecologically friendly. Where publications such as *The Green Consumer Guide* exhort consumers to wield their consumer power to force manufacturers to develop more ecologically-sensitive production processes, the bottom line is that consumption patterns need to be reduced rather than simply refocused.

Energy policy would again follow a similar hierarchy to resource and waste management. Conservation of energy becomes just as important as the consistently ignored and underfunded low-impact, renewable 'soft' energy options such as wind, sun and wave. Although these energy paths may not themselves generate enough to fulfil current demand, a reduction in consumption and conservation of energy would reduce this demand substantially. Measures such as energy and carbon taxes would be introduced to facilitate this change to a conservation society and the move from non-renewable fossil fuels to renewable sources.

This use of fiscal instruments to reduce the throughput of energy and raw materials is a key aspect of the green programme. The tax base itself would be altered in a move away from taxes on work and labour (for instance, income tax on the poor and national insurance), to be replaced by taxes on environmentally-damaging practices such as intensive energy usage, polluting industrial processes and the use of scarce raw materials. In this sense the tax system would focus on activities that ought to be discouraged from the standpoint of attaining a sustainable future. The polluter pays principle (PPP) becomes central.

The importance placed on the development of new renewable energy and pollution-control technologies challenges the frequent portrayal of greens as 'anti-technologists' or 'neo-Luddites'. There certainly are greens who espouse reactionary attitudes towards any form of technological development, but they tend to remain peripheral figures within the environmental movement. There is however a widespread recognition that technology tends to shape contemporary society, creating new wants and desires; its social and ethical impact often leading to a sharp separation of humanity from the natural world. Strong anthropocentric, Promethean attitudes of control and domination abound. For greens this attitude and relation to technology must be reversed.

Technology should be used appropriately in a manner which both aids the development of humanity in its widest, enlightened sense and has as little impact on the non-human world as possible. Nowhere is this more obvious than in the attitude of greens towards nuclear power, a focus of consistent campaigning by almost all sectors of the green movement. A potentially highly destructive energy source has been developed while low-impact alternatives have consistently been ignored.

The potential impact of nuclear technology is not simply a concern of the present generation - we are leaving future generations the legacy of dealing with the highly radioactive waste products of our energy generation. This is perhaps the most obvious example of both the present generation's over-optimistic belief in the promise of science to provide technological solutions to environmental problems and their failure to consider fully their obligations to the future. Inseparable to these considerations is the recognition that the nuclear energy and arms industries are inextricably linked. The idea of weapons of mass destruction that could obliterate all forms of life on the planet is simply anathema to green thinking.

Again, policies on agriculture have been developed in opposition to current unsustainable industrialised, intensive farming methods. The two core principles of green political thought combine to offer a strong critique of current practices. The mechanisation of farming and indiscriminate use of chemical fertilisers, pesticides and herbicides put high levels of stress on the land and pollute the soil, air and water courses. Wildlife and biodiversity is threatened through the loss of habitat and pollution and the introduction of monocultures, often promoted by the biotechnology industry. Intensive practices undermine humane animal husbandry. Although greens do not necessarily believe eating meat is wrong per se, they abhor contemporary factory-farming methods. Food safety, particularly in the light of BSE, use of animal hormones and GM foods, has become a crucial issue and organic farming, particularly at a community level, is held up as the way forward for agriculture.

Needs, Justice and Self-reliance

The policies discussed above are directly related to the two core insights of green political thought. The recognition of limits to growth provides the rationale for conservation of resources and the search for low-impact technology and energy sources. The desire for a humane and organic agricultural policy promotes a more sensitive relationship with non-human nature. The strong anthropocentric attitude underlying many industrial practices is questioned, particularly the potentially ecologically-devastating nuclear industry.

But it would be a disservice to the green platform if we imagined that green policies reflect a concern with only limits to growth and our ethical relations with non-human nature. Quite clearly the 'Statement of Core Principles' at the beginning of this case study is far broader. Justice and equality are key principles within a green economic, social and political framework grounded in the 'satisfaction of needs' rather than wants, a 'reconceptualisation of the nature of work' and a 'commitment to economic self-reliance'. As the 'Philosophical Basis' of the Green Party's *Manifesto for a Sustainable Society* states:

> As human beings, we all have the potential to live co-operatively and harmoniously with each other, and with reverence and respect for the complex web of life of which we are a part. Yet is has become increasingly obvious that the potential cannot be realised while basic human needs remain largely unmet. By basic needs we mean not only the physiological needs of food, water, air, shelter and sleep, but also psychological needs. These include the need for love, respect, autonomy, security, and meaningful activity within our communities.

The green programme is committed to a strong account of social justice and equity, based on the universal provision of needs and ecological security. Later in the Philosophical Basis, it is argued that 'a sense of insecurity will prevent individuals from acting in accordance with ecological constraints. A guarantee of security will not secure voluntary ecological behaviour, but it is a necessary precondition'.

This account of basic needs or security is tied in with a reconceptualisation of work and the economy. The green platform aims for a move away from the connection of meaningful work with paid employment. Outside the 'formal' wage-economy people engage in activities whose worth is undervalued in contemporary society - parenting, caring, community volunteering, etc. These activities of the informal economy are fundamental to both the economic and social spheres of life. Primarily to fulfil the conditions for social justice, greens argue that there should be a universal basic income scheme.

This would provide all citizens with a level of income sufficient to cover basic needs, replacing the current means-tested benefits system that often acts as a barrier to those who wish to take part-time employment or work on informal community-based activities. Critics question whether the incentive to work would remain (particularly given that the green vision is labour intensive) or whether the tax base can deliver such a scheme. This later issue is particularly problematic if economies are to be reduced in scale - there may simply not be enough money to re-distribute.

However, this is a long-term problem, and in the long term greens would expect our conception of money, work and community relations to have changed significantly. They point to local money and trading schemes, such as LETS (local exchange trading system), which have particular application in deprived areas, as offering the potential for more congenial community-based, not-for-profit forms of trading and organisation.

The green platform is firmly internationalist in its desire to ensure universal basic needs fulfilment and global justice. The reduction of conflict through aid, diplomacy and provision of appropriate technology and the end to the arms race and arms trade are fundamental policies. Greens are not however complete pacifists and there is a recognition that the armed forces and military intervention in a peacekeeping role may be necessary. Although the green programme has global pretensions, local self-reliance remains a fundamental principle.

The removal of debt and the restructuring of global economic institutions such as the World Bank, International Monetary Fund and the World Trade Organisation would aim to promote self-reliance. The transfer of aid and appropriate technology would promote the provision of basic needs through health care, education, family planning and self-sufficiency in food and energy. The idea of developing self-reliance is at the heart of the green position on global trade which would become the exception rather than the rule.

Shipping goods across the world is both polluting and inefficient. It also means that transnational corporations tend to have a high degree of influence on world markets, often leaving localities relatively powerless, particularly in the South.

This does not mean that more economically-developed countries do not have a duty to help such nations develop, but that this development should lead to nations and localities being able to provide for the basic needs of their population. Such an emphasis on self-reliance places greens in obvious opposition to the current trend towards global market liberalisation. Their position is often understood as a form of protectionism. However, it is protectionism of a particular kind - the development of local economies to provide for local needs.

This principle of localism means that politically the green programme calls for decentralisation and local democracy. Decisions should be made at the lowest appropriate level. Thus we find that greens are deeply concerned with the centralising tendency within the European Union and with the development of a single European currency. The EU is seen to be too remote from its citizens and dominated by economic interests. A democratic deficit exists. This does not mean that there is support for withdrawal from the EU; after all, it is seen as the appropriate political level for certain activities such as setting minimum ecological standards, safeguarding human and animal rights, redistributing resources within Europe and resolving disputes. However, decentralisation of political power wherever possible takes priority.

Finally, not only is prevention the order of the day when it comes to pollution and ecological destruction, but it is also the key principle of social policy. Currently, the majority of health services are focused on treating the symptoms of illness, with relatively few resources devoted to tackling the causes. Not only would more emphasis be placed on public health care, but preventative measures such as reducing pollution and traffic accidents, and ensuring food safety, are just as much aspects of health policy. The interconnectedness of different policy areas is obvious; ecologically-sensitive policy integration is at the heart of the green programme. Controversially, the green programme also advocates politically contentious policies towards crime and drugs. These policies have both libertarian and communitarian roots. For example, the decriminalisation of drug use is a libertarian policy; whereas criminal justice has a more communitarian focus, with an emphasis on rehabilitation, community service and restorative justice.

The green platform is far from single-issue and is developed from a broad range of principles. These principles are not purely 'ecological' and go beyond a concern about limits to growth and the value of non-human nature. Social justice based on the provision of basic needs, social policy that draws on liberal and communitarian principles, and economic and political decentralisation and democratisation are at the heart of a distinctive programme that sets the green programme apart from the policies of traditional political parties.

Index

A

Agracetus 128
AIDS virus 108, 131
Ames Test 138
Antibiotics 40, 118, 133, 135, 236
Applied genetics 79, 133, 134
Arms race 294
Artificial 52, 117, 122, 123, 128
Asilomar 141, 152

B

Bacillus thuringiensis 128, 226
Baconian legacy 120
Bacteria 20, 40, 102, 132, 133, 139, 148, 150, 204, 210, 224
Bioeffluent 148
Bioethicists 134
Biological utopia 120
Biologists 122, 140, 146, 152
Biosensors 133
Biotechnics 134

C

Cambridge 143, 152
Chemicals 2, 3, 8, 11, 12, 13, 15, 16, 19, 31, 35, 109, 127, 137, 138, 228
Civilization 76
Clones 124
Clostridium acetobatylicum 132
Commercial biotechnology 78
Commercialization 73, 222, 236, 238, 246
Consumers 20, 39, 183, 216, 224, 236, 240, 290
Containment 131, 141, 142, 143, 144, 146, 148, 156
Controversy over 155
Criticism from 144, 146

D

Debate over 98, 152, 153, 155, 250
Deliberate release 135, 149, 153, 155, 156
Deterministic 102
Distinction 80, 83, 102. 109, 148, 150, 185, 271

E

Earth first 156
Ecological society 271
Ecologists 60, 154, 155, 156, 188, 248, 267
Ecology 74, 79, 117, 132, 133, 138, 150, 155, 255, 259, 261, 265, 266, 273

Effectiveness 43, 177, 182, 188
Environmental action 157, 167
Environmental release 78, 136, 156
Environmentalism 259, 271, 272, 281
Environmentalists 134, 148, 155, 159, 257, 259, 260, 275, 281
Erwinia herbicola 151
Ethylene oxide 15
Eutrophication 128
Evolution 69, 84, 85, 99, 102, 104, 105, 107, 108, 109, 111, 114, 115, 119, 123, 125, 203, 205, 206, 208, 211

F

Factory farming 204
Fantasy 110
Farmers 41, 45, 48, 118, 197, 224, 226, 227, 231, 232, 233, 234, 235, 242
Fermentation 20, 28, 40, 146, 148
Field-testing 140, 149, 150, 154
Food and drug 75
For synthetic chemicals 137
Forces of production 253, 262, 263
Frances sharples 156
Friends of the earth 152, 153

G

Gene therapy 81, 134, 197
General electric 126
Genesis 117
Gentechnics 125, 136
Germline 124, 154
Goodrich 127
Green revolution 224

H

Hazards 2, 3, 8, 16, 30, 39, 61, 74, 75, 133, 138, 155, 210, 211, 212, 216, 218
Herbicide 18, 127, 130, 227, 233, 234, 241, 242
Herbicide-tolerant 233, 234, 241, 242
Historical 64, 68, 82, 140, 220, 252, 255, 269
Hormone 57, 203, 210, 215, 217
Human existence 198, 201
Human gene therapy 81, 134
Human genetic engineering 83
Hybrid 132, 207, 235
Hybridization 118, 123, 135

I

Identification 60, 62, 86, 137, 139, 259, 273
Incrementalism 74
Insecticide 226

J

Jeremy rifkin 115

L

Largescale 146
Legislation 16, 23, 50, 74, 141, 143, 144
Liberation 87
Litigation 75, 151, 153, 154

M

Mammals 44, 52, 58, 206, 230
Manipulation 124, 154, 201, 204, 205, 206, 207, 210, 211, 236, 283
Martin alexander 156
Massachusetts 126, 143, 152

Mechanization 202
Methodology 124, 138
Microbial 128, 132, 133, 138
Microbiologically 132
Microbiology 126
Military 64, 92, 97, 98, 294
Molecular genetics 123, 125, 141, 205
Monsanto 18, 26, 119, 127, 133, 200, 202
Mycogen Corporation 127

N

National academy of sciences 130, 154
National institutes of health 140
National science foundation 38
Novelty 82, 151, 206, 257
Nuclear technology 139, 209, 291
Nurture 269

P

Paradigm 79, 121, 146, 207
Patent claim 126
Patenting 75, 198, 199, 200, 212
Pesticides 15, 46, 63, 126, 127, 202, 219, 220, 225, 226, 227, 230, 241, 245, 292
Pharmaceuticals 15, 35
Philosophical 40, 153, 195
Plant breeder 123
Political ideology 273, 274
Premanufacture 39
Pseudomonas 126, 133, 135, 151
Public sector 75

R

R. H. Zaugg 133
Reactionism 75
Recombinant DNA 81, 82, 118, 135, 140, 149
Revolution 74, 75, 80, 82, 115, 119, 126, 134, 136, 140, 201, 263

S

Scarcity of 130
Seed 41, 42, 128, 129, 198, 199, 214, 225
Sierra club 152
Social governance 74, 76, 77
Social ownership 75, 76
Social problems 254, 266
Stanford University 140, 149
Syndrome 64

T

Task force 154
Technology assessment 78, 145
Technology assessment 74, 77, 78, 79, 80, 81, 82, 83
The new atlantis 120
Thiobacillus ferroxidans 132, 133
Toxic substances control act 133
Toxicological 138
Traditional orientation 79
Transgenic 81, 124, 131
Treatment 14, 31, 51, 108, 126, 128, 200, 217, 258

U

Universities 1, 25, 37, 142
University of amsterdam 130
University of california at berkeley 151

V

Vietnam war 63

Voluntary 61, 142, 144, 145, 146, 147, 148, 164, 288, 293

W

Working group 189, 190, 191

X

Xanthomonas campestris 133

Y

Young 10, 142, 223